THE C. S. CARD IRON WORKS CO.

MINE HAULAGE AND HANDLING EQUIPMENT

Catalog

Designers and Manufacturers

Established 1892

No. 40

C. S. CARD IRON WORKS CO.

Section "A"—Mine Car Wheels and Trucks

Section "B"—Coal Mine Cars

Section "C"—Ore and Industrial Cars

Section "D"—Rope Haulage Equipment, Rollers, Sheaves, etc.

Section "E"—Tipple Equipment, Dumps, Cages, Screens, etc.

Section "F"—Track Equipment, Frogs, Switches, Crossings, etc.

Cable and Radio Address

"CARDIRON" DENVER

Main Office and Works

2501 WEST SIXTEENTH AVE., DENVER, COLORADO

©2008-2010 PERISCOPE FILM LLC
ALL RIGHTS RESERVED
ISBN#978-1-935700-28-9
WWW.PERISCOPEFILM.COM

INDEX

INDEX—Continued

SALES CONDITIONS

Quotations and Orders are Given and Accepted Subject to the Following Conditions:

All quotations and prices are made for immediate acceptance, and are subject to change in price without notice, unless otherwise agreed upon in writing.

Quotations made on material from stock are subject to prior sale.

All quotations, deliveries and agreements are made contingent on strikes, accidents or other unavoidable delays, over which we have no control.

All agreements and specifications must be made in writing and accepted by an officer of this Company or an authorized representative. No verbal agreements or understandings will be recognized.

All quotations made f.o.b. destination are based on freight rates prevailing on date quotation is made, and rate quoted us by Transportation Company. Any error, or change in rate made prior to shipment by Transportation Company must be absorbed by purchaser.

Agreements regarding time of delivery, unless otherwise specified, will be figured from date of final settlement of all details, whether technical or otherwise.

Orders entered will not be subject to cancellation by purchaser without our consent, and then only upon terms that will indemnify us against all loss.

Shipments made at purchaser's risk. Transportation Company's receipt in good order constitutes delivery, after which our responsibility ceases.

Weights given are only approximate, and are not binding.

Cuts shown in this catalogue or referred to in specifications, are only to show general design—we reserve the right to make such changes as we deem necessary, without notice.

Every care is taken by us to insure first-class material and workmanship, but beyond replacing f.o.b. our works any material, accepted by us as defective, we will in no case be responsible for any expense or consequences due to such defects. The acceptance of material by purchaser is understood to constitute a waiver of all claims for damages on account of delay.

By the terms "gauge" or "track gauge," the distance between the inside heads of the rails is always understood.

A

SECTION "A"
CATALOG NO. 40

MINE CAR WHEELS
TRUCKS AND PARTS

The C.S. Card Iron Works Co.
Denver, Colorado

A

"CARD" WHEELS AND TRUCKS

On the following pages we illustrate and describe various wheels and trucks we regularly manufacture, together with a few of special design.

Our standard design trucks have been developed over a long period and have been thoroughly tested. They will meet most conditions encountered in the mine haulage field.

Special trucks to meet special or unusual conditions can be developed and furnished. Our long experience in this field is at your disposal.

With our modern foundry making a specialty of chilled castings, we are in position to furnish you with a high quality wheel, the equal of any on the market.

All truck parts are made from materials selected for the work they are to do and which our long experience indicates are best adapted for the several purposes. Axles of high carbon steel; rollers, bushings, cotters, etc., are all of special analysis steel, made to our specifications.

"CARD" Semi-Steel Heat-Treated Wheels are a semi-steel casting, heat-treated, with a deep even chill, and a close-grained, tough spoke or plate and hub. They stand the roughest service.

You may wish us to make a recommendation as to the size of the axle and weight of the wheel to use, or the design of the truck, for your conditions. If so, furnish us such information as the gross load of car and material; type of haulage; maximum speed; length of haul; track conditions such as weight of rail, curvatures, etc. All of these factors should be given consideration in selecting a truck from which you can expect to get maximum satisfaction.

We can furnish trucks to mount on the car frames of your present equipment. If you will furnish us a dimensioned sketch of your car frame, we will gladly send you a drawing showing how it can be done. (See page 31.)

In ordering standard trucks or asking for quotations, please furnish us the following information:

Style of truck, Page

Diameter of wheel (measured at the tread)

Size of the axle and whether round or square

Track gauge (distance between heads of rail)

If a round axle, specify the pedestal style (See page 30).

By the term Gauge or Track Gauge it is understood that the distance between the inside heads of rails is meant, thus:

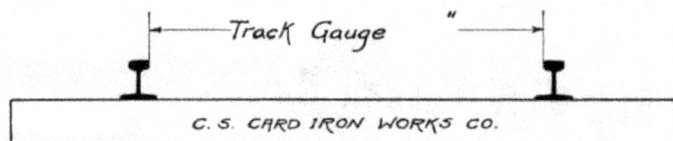

We make the necessary allowance for clearance.

"CARD" IMPROVED ROLLER BEARING WHEEL

A thoroughly Time-Tested, Mine-Tested wheel developed from our original solid closed cap wheel, which was the pioneer roller bearing wheel with a solid cap.

From the illustration you can see the simplicity of the wheel and its few parts, which have made it trouble-proof and a favorite with the trade.

The solid cap, a part of the wheel casting, cannot be lost. There are no bolts, screws, etc., to become loose, lost or otherwise cause trouble.

The solid cap permits the end thrust to be taken from the inside of the hub and places it between the end of the axle and the boss on the inside of the wheel cap, making a lighter running truck, especially on curves.

The wheel is held on the axle by a special analysis chrome alloy steel cotter; a most simple and positive fastening that will last the life of the wheel.

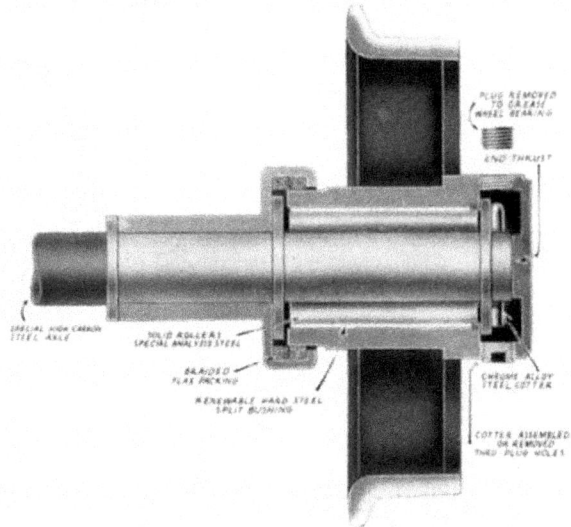

The openings for the insertion and removal of the cotter are closed with countersunk plugs, which on account of taper and fine thread do not jar out and are easily removed when necessary. The wheel is lubricated through these same openings.

We recommend lubricating with grease or hard oil. The frequency of lubrication depends on the service. For ordinary service, we recommend lubrication at three-month intervals. The wheel will run longer than this on one application.

The solid dirt-proof cap in the front of the wheel and the braided flax packing in the pedestal bell at the rear of the hub make a seal, retaining the lubricant and excluding dirt from the bearings, thus increasing their life.

The wheel is designed with a liberal bearing length, located so that the load and thrust are equally distributed and wear is reduced to a minimum.

A special analysis steel split bushing takes the wear in the wheel hub and is renewable when necessary.

The straight, solid roller bearings are of a special analysis steel, and are mounted without a cage, thus eliminating another source of trouble. They are proportioned so as to properly carry the load in connection with the high carbon steel axles with which they are used.

The wheel casting is semi-steel, made from a special iron mixture carrying a high percentage of steel; heat-treated; and has a deep evenly chilled tread and close-grained strong tough spoke or plate and hub.

The grey iron pedestals which carry the axles and attach to the car frame are liberally proportioned. They can be designed to attach to almost any car frame.

The wheel has four cardinal points, namely:—its simplicity of design with few working and wearing parts; its dirt-proof and economical lubrication; its easy running; and its thoroughly demonstrated economy over years of actual mine service.

"CARD" ROLLER BEARING TRUCKS

ROUND AXLES WITH PACKED PEDESTALS

This most popular design "CARD" Improved Roller Bearing Truck is used by the thousands, and has proven itself under many different field operating conditions.

For a full description of the wheel refer to page 3 in this section.

The round axle being free to turn in the pedestal, all the axle wear is taken evenly instead of on one side only as is the case with square axles. The axles and roller bearings being of special analysis steel and the bearings of liberal length and properly located in connection with the load, wear is negligible.

The pedestals for attaching the truck to car frames can be had in many styles and to meet almost any condition. See page 30 for our standard pedestals and page 31 showing a number of pedestal applications. If you will furnish us with a dimensioned sketch of your car frame to which you wish to attach this truck, we will gladly submit a drawing showing how it can be done.

The upper illustration shows the truck with a Style "A" pedestal; the lower illustration with a Style "C" pedestal.

These trucks are shipped grease packed ready to operate.

Trucks can be furnished with the wheels of either spoke or plate pattern.

See page 5 for price list and weights of this truck.

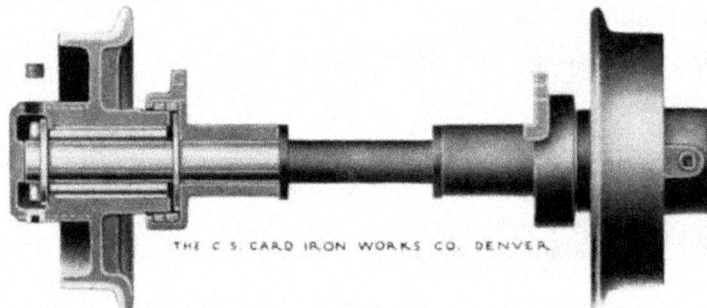

This truck, with Style "C" pedestals, is regularly used on our Type "Z" ore cars. Trucks are carried in stock for immediate shipment in the following sizes:

10" Plate Wheels—1½" Axles—18" Gauge
10" Plate Wheels—1¾" Axles—18" and 24" Gauge
12" Plate Wheels—1¾" Axles—18" and 24" Gauge
12" Plate Wheels—2" Axles—18" and 24" Gauge

Dimension sheets covering these trucks are available on request.

"CARD" ROLLER BEARING TRUCKS
ROUND AXLES WITH PACKED PEDESTALS

Prices and weights listed below cover one truck. A truck consists of four wheels (spoke or plate pattern) with bearings, four pedestals and two axles, assembled and grease packed, ready to attach to car.

When ordering or inquiring for prices, please specify the wheel diameter, the axle size, the track gauge, and the pedestal style. State whether spoke or plate pattern wheels are desired.

Lists cover a truck with standard Style "A" and "B" pedestals; Style "C" and "D" pedestals carry extras. See page 30 for pedestal styles.

Wheel Diam. And Weight	Axle Size	PRICE PER TRUCK					Add for Type 'C' Pedestal	Add for Type 'D' Pedestal	WEIGHT (Lbs.) PER TRUCK					Add for Type 'D' Pedestal
		18″	24″	30″	36″	42″			18″	24″	30″	36″	42″	
8″–C	1½″	$26.66	$27.00	$27.34	$27.68	$28.02	$0.72	$1.26	225	231	237	243	249	8
	1¾″	29.39	29.85	30.31	30.77	31.23	0.72	1.30	256	264	272	280	288	9
10″–B	1½″	27.59	27.93	28.27	28.61	28.95	0.72	1.26	261	267	273	279	285	8
	1¾″	30.33	30.79	31.25	31.71	32.17	0.72	1.30	292	300	308	316	324	9
	2″	33.33	33.93	34.53	35.13	35.73	0.72	1.31	345	356	367	378	389	10
10″–C	2¼″	37.41	38.19	38.97	39.75	40.53	0.75	1.39	397	411	425	439	453	12
	2½″	42.27	43.23	44.19	45.15	46.11	0.75	1.57	458	475	492	509	526	20
12″–B	1½″	29.15	29.49	29.83	30.17	30.51	0.72	1.26	317	323	329	335	341	8
	1¾″	31.89	32.35	32.81	33.27	33.73	0.72	1.30	348	356	364	372	380	9
	2″	34.94	35.54	36.14	36.74	37.34	0.72	1.31	401	412	423	434	445	10
12″–C	2¼″	39.08	39.86	40.64	41.42	42.20	0.75	1.39	457	471	485	499	513	12
	2½″	43.88	44.84	45.80	46.76	47.72	0.75	1.57	518	535	552	569	586	20
12″–D	2¾″	48.49	49.63	50.77	51.91	53.05	0.79	1.72	602	622	642	662	682	24
14″–B	1½″	30.61	30.95	31.29	31.63	31.97	0.72	1.26	365	371	377	383	389	8
	1¾″	33.40	33.86	34.32	34.78	35.24	0.72	1.30	396	404	412	420	428	9
14″–C	2″	37.02	37.62	38.22	38.82	39.42	0.72	1.31	473	484	495	506	517	10
	2¼″	40.74	41.52	42.30	43.08	43.86	0.75	1.39	509	523	537	551	565	12
14″–D	2½″	45.91	46.87	47.83	48.79	49.75	0.75	1.57	586	603	620	637	654	20
14″–E	2¾″	50.93	52.07	53.21	54.35	55.49	0.79	1.72	690	710	730	750	770	24
14″–F	3″	58.30	59.68	61.06	62.44	63.82	0.79	1.81	798	822	846	870	894	28
16″–C	1¾″	35.63	36.09	36.55	37.01	37.47	0.72	1.30	485	493	501	509	517	9
	2″	38.63	39.23	39.83	40.43	41.03	0.72	1.31	537	548	559	570	581	10
	2¼″	42.30	43.08	43.86	44.64	45.42	0.75	1.39	573	587	601	615	629	12
16″–D	2½″	47.57	48.53	49.49	50.45	51.41	0.75	1.57	650	667	684	701	718	20
16″–E	2¾″	52.75	53.89	55.03	56.17	57.31	0.79	1.72	762	782	802	822	842	24
16″–F	3″	59.81	61.19	62.57	63.95	65.33	0.79	1.81	854	878	902	926	950	28
18″–C	2″	40.81	41.41	42.01	42.61	43.21	0.72	1.31	605	616	627	638	649	10
	2¼″	44.49	45.27	46.05	46.83	47.61	0.75	1.39	641	655	669	683	697	12
18″–D	2½″	50.07	51.03	51.99	52.95	53.91	0.75	1.57	734	751	768	785	802	20
18″–E	2¾″	55.30	56.44	57.58	58.72	59.86	0.79	1.72	846	866	886	906	926	24
18″–F	3″	62.77	64.15	65.53	66.91	68.29	0.79	1.81	958	982	1006	1030	1054	28

8″ Diameter wheels in plate pattern only.
10″ Diameter wheels will be furnished in plate pattern unless otherwise specified.
See page 20 for information on wheel weights, wheel treads, etc.
See page 29 for axle clips and cross channels.

"CARD" TIMKEN ROLLER BEARING TRUCKS

ROUND AXLES WITH PLAIN PEDESTALS

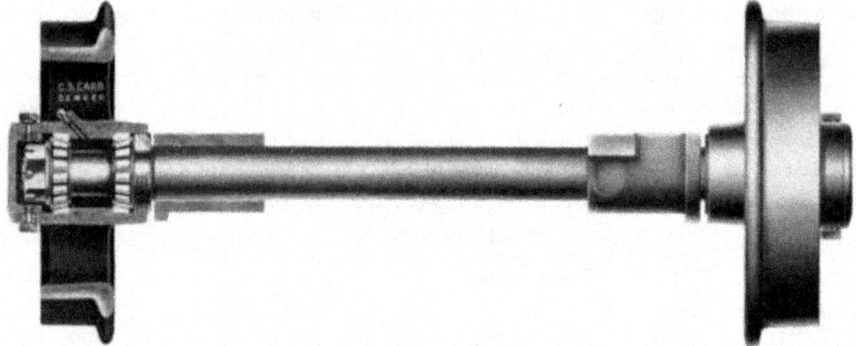

A very popular easy running truck, with Timken bearings mounted in the wheels, and using round high carbon steel axles carried in plain pedestals.

The design follows the recommended practices of the Timken Roller Bearing Company.

We are exceptionally well equipped with machines, jigs, etc., for accurately producing this truck. To this, and our long experience in the quantity production of this style truck, we attribute in large part the very satisfactory results obtained by the operators of "CARD"-Timken trucks.

The wheel casting is semi-steel, made from a special iron mixture carrying a high percentage of steel; heat-treated; and has a deep even chill and close-grained strong, tough plate or spokes and hub.

Axles are a special high carbon steel.

The pedestals for attaching the truck to the car frame can be had in many types and to meet almost any condition. See page 30 for our standard pedestals and page 31 showing a number of applications.

If you will furnish us with a dimensioned sketch of your car frame to which you wish to attach this truck, we will gladly submit a drawing showing how it can be done.

See page 7 for price list and weights of this truck.

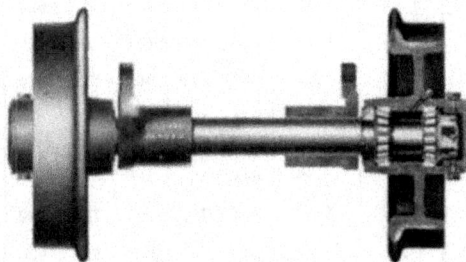

Our standard type "Z" ore cars can be equipped with a truck of this design, with Style "C" pedestals and these trucks are carried in stock for immediate shipment in the following sizes:

10" Plate Wheels—1¾" Axles—18" and 24" Gauge.
12" Plate Wheels—1¾" Axles—18" and 24" Gauge.
12" Plate Wheels—2" Axles—18" and 24" Gauge.

Dimension sheets covering these trucks are available on request.

The design of truck illustrated will meet most of the conditions found in the mine car field. However, different styles of bearings, seals, mountings, etc., can be furnished where necessary or advisable.

"CARD" TIMKEN ROLLER BEARING TRUCKS
ROUND AXLES WITH PLAIN PEDESTALS

Prices and weights listed below cover one truck. A truck consists of four wheels (spoke or plate pattern) with cast iron caps, Timken roller bearings, two axles, and four pedestals, assembled and grease packed, ready to attach to car.

When ordering or inquiring for prices, please specify the wheel diameter, the axle size, the bearing set number, the track gauge, and the pedestal style. State whether spoke or plate pattern wheels are desired.

Lists cover a truck with standard Style "A" and "B" pedestals; Style "C" and "D" pedestals carry extras. See page 30 for pedestal styles.

Wheel Diam. and Weight	Axle Size *	Bearing set No. **	PRICE PER TRUCK					Add for Type 'C' Pedestal	Add for Type 'D' Pedestal	WEIGHT (Lbs.) PER TRUCK					Add for 'D'
			18"	24"	30"	36"	42"			18"	24"	30"	36"	42"	
8"–C	1¾"	E-2701	$37.09	$37.55	$38.01	$38.47	$38.93	$0.72	$1.35	238	246	254	262	270	12
10"–B	1¾"	E-2701	37.66	38.12	38.58	39.04	39.50	0.72	1.35	262	270	278	286	294	12
	2"	E-2702	42.77	43.37	43.97	44.57	45.17	0.72	1.40	295	306	317	328	339	14
10"–C	2¼"	E-2703	47.28	48.06	48.84	49.62	50.40	0.75	1.48	343	357	371	385	399	16
	2½"	E-2704	52.62	53.58	54.54	55.50	56.46	0.75	1.66	396	413	430	447	464	24
12"–B	1¾"	E-2701	39.38	39.84	40.30	40.76	41.22	0.72	1.35	326	334	342	350	358	12
	2"	E-2702	44.17	44.77	45.37	45.97	46.57	0.72	1.40	347	358	369	380	391	14
12"–C	2¼"	E-2703	48.68	49.46	50.24	51.02	51.80	0.75	1.48	395	409	423	437	451	16
	2½"	E-2704	53.81	54.77	55.73	56.69	57.65	0.75	1.66	436	453	470	487	504	24
12"–D	2¾"	E-2705	58.94	60.08	61.22	62.36	63.50	0.79	1.81	505	525	545	565	585	28
14"–B	1¾"	E-2701	40.73	41.19	41.65	42.11	42.57	0.72	1.35	366	374	382	390	398	12
14"–C	2"	E-2702	45.99	46.59	47.19	47.79	48.39	0.72	1.40	407	418	429	440	451	14
	2¼"	E-2703	50.14	50.92	51.70	52.48	53.26	0.75	1.48	439	453	467	481	495	16
14"–D	2½"	E-2704	55.69	56.65	57.61	58.57	59.53	0.75	1.66	500	517	534	551	568	24
14"–E	2¾"	E-2705	61.12	62.26	63.40	64.54	65.68	0.79	1.81	581	601	621	641	661	28
14"–F	3"	E-2707	79.14	80.52	81.90	83.28	84.66	0.79	1.90	663	687	711	735	759	32
16"–C	1¾"	E-2701	49.40	49.86	50.32	50.78	51.24	0.72	1.35	442	450	458	466	474	12
	2"	E-2702	47.76	48.36	48.96	49.56	50.16	0.72	1.40	475	486	497	508	519	14
	2¼"	E-2703	51.91	52.69	53.47	54.25	55.03	0.75	1.48	507	521	535	549	563	16
16"–D	2½"	E-2704	57.45	58.41	59.37	60.33	61.29	0.75	1.66	568	585	602	619	636	24
16"–E	2¾"	E-2705	62.79	63.93	65.07	66.21	67.35	0.79	1.81	645	665	685	705	725	28
16"–F	3"	E-2707	80.81	82.19	83.57	84.95	86.33	0.79	1.90	727	751	775	799	823	32
18"–C	2"	E-2702	49.79	50.39	50.99	51.59	52.19	0.72	1.40	539	550	561	572	583	14
	2¼"	E-2703	53.93	54.71	55.49	56.27	57.05	0.75	1.48	571	585	599	613	627	16
18"–D	2½"	E-2704	59.48	60.44	61.40	62.36	63.32	0.75	1.66	632	649	666	683	700	24
18"–E	2¾"	E-2705	64.97	66.11	67.25	68.39	69.53	0.79	1.81	717	737	757	777	797	28
18"–F	3"	E-2707	83.30	84.68	86.06	87.44	88.82	0.79	1.90	811	835	859	883	907	32

*—Rough axle diameter. Finished diameter at the bearings is ¼" less.
**—Bearings carried in stock for immediate shipment.

8" Diameter wheels in plate pattern only.
10" Diameter wheels will be furnished in plate pattern unless otherwise specified.
See page 20 for information on wheel weights, wheel treads, etc.
See page 29 for axle clips and cross channels.

"CARD" PATENTED ROLLER BEARING
SPRING DRAWBAR TRUCKS

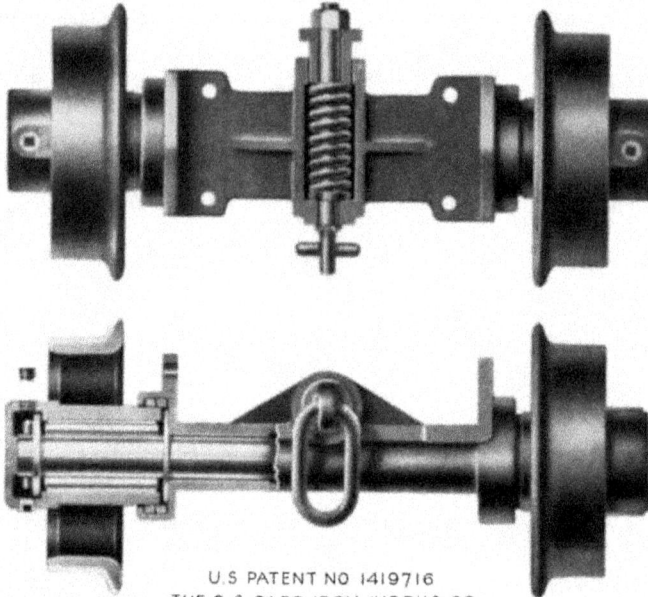

U.S PATENT NO 1419716
THE C. S. CARD IRON WORKS CO.

This truck was developed and is recommended for use on small ore cars in long trains with fast motor haulage. By eliminating the jerking and shocks on motor and cars, it reduces haulage and repair costs.

The pedestal casting encloses the drawbar spring as well as the axle, making a one-piece unit and transmitting the drawbar pull direct to the truck and frame.

Each car obtains the full benefit of two springs in starting.

In starting a train, the springs gradually take the pull until partially compressed, when the drawbar pull is transferred from the spring to the pedestal.

By this method the train is easily started without a sudden jerk, and the springs cannot be compressed enough to cause breakage by the coils riding each other.

The illustration shows a roller bearing wheel as described on page 3.

The truck can also be furnished with the CARD Timken roller bearing wheel shown on page 6.

The drawbar and pedestal can be designed to meet any requirements at a slight increase over the standard list below.

Drawbar hitchings can be made to mix with your present equipment.

If you will furnish a drawing or dimensioned sketch of your present car equipment, we will submit a drawing showing what can be done.

This truck is regularly furnished on our standard "S" Type Ore Cars.

Prices and weights listed cover a truck which consists of four wheels with bearings, two axles, two pedestals with springs, drawbars and couplings, assembled, grease packed and ready to attach to car.

With "CARD" Improved Roller Bearing Wheels

Wheel Diam. And Weight	Axle Size	PRICE PER TRUCK Track Gauge		WEIGHT (Lbs.) PER TRUCK Track Gauge	
		18"	24"	18"	24"
10"–B	1¾"	$53.90	$54.64	384	405
	2"	59.31	60.18	440	463
10"–C	2¼"	63.93	65.00	490	515
12"–B	1¾"	55.87	56.62	456	477
	2"	61.18	62.05	508	531
12"–C	2¼"	65.91	66.98	562	587
14"–B	1¾"	57.90	58.64	520	541
14"–C	2"	63.67	64.55	592	615
	2¼"	68.04	69.11	630	655

With "CARD" Timken Roller Bearing Wheels

Wheel Diam. And Weight	Axle Size	PRICE PER TRUCK Track Gauge		WEIGHT (Lbs.) PER TRUCK Track Gauge	
		18"	24"	18"	24"
10"–B	1¾"	$56.52	$57.27	370	390
	2"	63.27	64.16	405	427
10"–C	2¼"	69.63	70.69	456	482
12"–B	1¾"	58.65	59.40	450	470
	2"	64.99	65.87	469	491
12"–C	2¼"	71.45	72.51	524	550
14"–B	1¾"	59.64	60.39	470	490
14"–C	2"	67.23	68.11	541	563
	2¼"	73.16	74.23	576	602

See page 20 for information on wheel weights, wheel treads, etc.

"CARD" IMPROVED PLAIN BEARING WHEEL

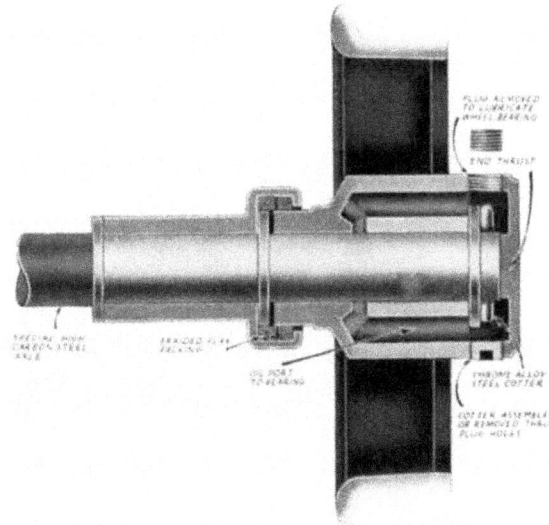

For those operators who desire a plain bearing wheel, we recommend the wheel shown above. It incorporates all of the important features of our improved roller bearing wheel, except for the bearings.

From the illustration you will see the simplicity of the design. The solid cap, a part of the wheel casting, eliminates all trouble from loose caps and their fastenings. It also permits the end thrust to be taken from the inside of the hub and placed on the wheel cap and end of the axle, making an easier running truck, especially on curves.

The openings for the removal and insertion of the cotter are closed with countersunk plugs which on account of taper and fine thread do not jar out and are easily removed when necessary. The wheel is lubricated through these same openings.

No lubricant can escape through the solid cap of the wheel, and the braided flax packing at the rear helps retain it. This also makes an effective seal against the entrance of dirt to the bearings, greatly increasing the life.

The wheel is designed with a liberal length of bearing, all of which is well lubricated from the cap and through the enclosed oil ports. The bearing is properly located to carry the load and thrusts.

The wheel is held on the axle by a special analysis alloy steel cotter, a most simple and positive fastening that will last the life of the wheel.

The wheel casting is semi-steel, made from a special iron mixture carrying a high percentage of steel; heat-treated; and has a deep even chill and close-grained tough strong spokes or plate and hubs, easily machined.

The axle is a high carbon steel.

The grey iron pedestals that carry the axle are liberally proportioned and can be furnished to meet almost any condition. See pages 30 and 31.

This wheel is used in different truck combinations on the following pages.

"CARD" PLAIN BEARING TRUCKS

ROUND AXLES WITH PACKED PEDESTALS

This is the most popular design of plain bearing truck we make. We recommend this truck to those who prefer to use a plain bearing wheel.

Refer to page 9 of this section for a full description of the wheel.

The round axle, being free to turn in the pedestals, wears evenly instead of on one side only, as does a square axle.

The axle being a special analysis high carbon steel and the wheel bearing of close-grained metal and of liberal length, with proper lubrication, the wear on both is reduced to a minimum.

The pedestals for attaching the truck to car frames can be had in many styles and to meet almost any condition. See page 30 for standard pedestals and page 31 showing a number of pedestal applications. If you will furnish us with a dimensioned sketch of your car frame to which you wish to attach this truck, we will gladly submit a drawing showing how it can be done.

The truck illustration above shows a style "A" pedestal as is ordinarily used on a wood bottom coal car.

Trucks can be furnished with the wheels of either spoke or plate pattern.

See page 11 for price list and weights of this truck.

"CARD" PLAIN BEARING TRUCKS
ROUND AXLES WITH PACKED PEDESTALS

Prices and weights listed below cover one truck. A truck consists of four wheels (spoke or plate pattern), four pedestals and two axles, assembled and ready to attach to car.

When ordering or inquiring for prices, please specify the wheel diameter, the axle size, the track gauge, and the pedestal style. State whether spoke or plate pattern wheels are desired.

Lists cover a truck with standard "A" and "B" pedestals; style "C" and "D" pedestals carry extras. See page 30 for pedestal styles.

Wheel Diam. And Weight	Axle Size	PRICE PER TRUCK					Add for Type 'C' Pedestal	Add for Type 'D' Pedestal	WEIGHT (Lbs.) PER TRUCK					Add for Type 'D' Pedestal
		Track Gauge							Track Gauge					
		18"	24"	30"	36"	42"			18"	24"	30"	36"	42"	
8"–C	1½"	$25.62	$25.96	$26.30	$26.64	$26.98	$ .72	$1.26	187	193	199	205	211	8
	1¾"	27.52	27.98	28.44	28.90	29.36	.72	1.30	216	224	232	240	248	9
10"–B	1½"	26.81	27.15	27.49	27.83	28.17	.72	1.26	239	245	251	257	263	8
	1¾"	28.77	29.23	29.69	30.15	30.61	.72	1.30	268	276	284	292	300	9
	2"	30.23	30.83	31.43	32.03	32.63	.72	1.31	300	311	322	333	344	10
10"–C	2¼"	33.09	33.87	34.65	35.43	36.21	.75	1.39	339	353	367	381	395	12
	2½"	36.51	37.47	38.43	39.39	40.35	.75	1.57	391	408	425	442	459	20
12"–B	1½"	28.27	28.61	28.95	29.29	29.63	.72	1.26	291	297	303	309	315	8
	1¾"	30.28	30.74	31.20	31.66	32.12	.72	1.30	320	328	336	344	352	9
	2"	32.00	32.60	33.20	33.80	34.40	.72	1.31	364	375	386	397	408	10
12"–C	2¼"	35.01	35.79	36.57	37.35	38.13	.75	1.39	411	425	439	453	467	12
	2½"	38.49	39.45	40.41	41.37	42.33	.75	1.57	463	480	497	514	531	20
12"–D	2¾"	41.65	42.79	43.93	45.07	46.21	.79	1.72	528	548	568	588	608	24
14"–B	1½"	29.99	30.33	30.67	31.01	31.35	.72	1.26	347	353	359	365	371	8
	1¾"	31.94	32.40	32.86	33.32	33.78	.72	1.30	376	384	392	400	408	9
14"–C	2"	33.66	34.26	34.86	35.46	36.06	.72	1.31	420	431	442	453	464	10
	2¼"	36.36	37.14	37.92	38.70	39.48	.75	1.39	451	465	479	493	507	12
14"–D	2½"	40.36	41.32	42.28	43.24	44.20	.75	1.57	527	544	561	578	595	20
14"–E	2¾"	43.63	44.77	45.91	47.05	48.19	.79	1.72	600	620	640	660	680	24
14"–F	3"	47.08	48.46	49.84	51.22	52.60	.79	1.81	672	696	720	744	768	28
16"–C	1¾"	33.86	34.32	34.78	35.24	35.70	.72	1.30	452	460	468	476	484	9
	2"	35.17	35.77	36.37	36.97	37.57	.72	1.31	476	487	498	509	520	10
	2¼"	37.82	38.60	39.38	40.16	40.94	.75	1.39	507	521	535	549	563	12
16"–D	2½"	42.39	43.35	44.31	45.27	46.23	.75	1.57	607	624	641	658	675	20
16"–E	2¾"	45.81	46.95	48.09	49.23	50.37	.79	1.72	680	700	720	740	760	24
16"–F	3"	49.16	50.54	51.92	53.30	54.68	.79	1.81	752	776	800	824	848	28
18"–C	2"	37.10	37.70	38.30	38.90	39.50	.72	1.31	536	547	558	569	580	10
	2¼"	39.79	40.57	41.35	42.13	42.91	.75	1.39	567	581	595	609	623	12
18"–D	2½"	44.26	45.22	46.18	47.14	48.10	.75	1.57	663	680	697	714	731	20
18"–E	2¾"	47.79	48.93	50.07	51.21	52.35	.79	1.72	744	764	784	804	824	24
18"–F	3"	51.45	52.83	54.21	55.59	56.97	.79	1.81	828	852	876	900	924	28

8" Diameter wheels in plate pattern only.
10" Diameter wheels will be furnished in plate pattern unless otherwise specified.
See page 20 for information on wheel weights, wheel treads, etc.
See page 29 for axle clips and cross channels.

"CARD" PLAIN BEARING TRUCKS

SOLID CAP WHEELS—SQUARE AXLES—NO PEDESTALS

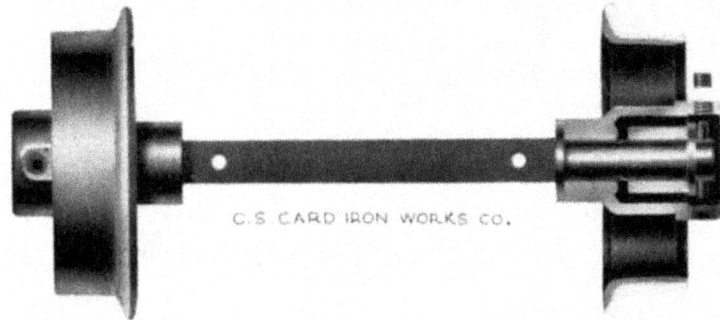

This is a combination of our standard Solid Cap plain bearing wheel used on a square axle with the rear of the hub counter-bored for the axle washer, but without pedestals.

The advantages of a solid cap are obvious; no lost caps or parts to replace, and any oil placed in the cap must pass through the bearing.

A light lubricant is recommended and the wheel is oiled through the hole in the center of the cap.

This truck will be found satisfactory where the haul is short and at slow speed, and initial cost is considered.

The truck is equipped with CARD'S semi-steel heat-treated wheels of either plate or spoke pattern.

The wheels have a liberal length of bearing which increases the life materially.

Axles are a special analysis high carbon steel—to better resist wear.

Alloy steel cotters, that will last the life of the wheel, are used to retain the wheel on the axle.

The illustration shows a Style No. 6 axle with holes for bolting the truck to a car frame.

Can be furnished with any style of axle shown on page 28.

For a larger view of the wheel only, see page 9, which very closely approximates the one used on this truck.

See page 13 for price list and weights of this truck.

"CARD" PLAIN BEARING TRUCKS

SOLID CAP WHEELS—SQUARE AXLES—NO PEDESTALS

Prices and weights listed below cover one truck. A truck consists of four wheels (spoke or plate pattern) and two axles, assembled and ready to attach to car.

When ordering or inquiring for prices, please specify the wheel diameter, the axle size and style number, and the track gauge. State whether spoke or plate pattern wheels are desired.

Lists cover a truck with style No. 1 axles; other styles of axles carry extras. See page 28 for axle styles and numbers.

Wheel Diam. and Weight	Axle Size	PRICE PER TRUCK					Add for Style #2 Axles	Add for Style #6 Axles	WEIGHT (Lbs.) PER TRUCK				
		Track Gauge							Track Gauge				
		18"	24"	30"	36"	42"			18"	24"	30"	36"	42"
8"–C	1¼"	$19.41	$19.73	$20.05	$20.37	$20.69	$0.26	$0.22	118	123	128	133	138
	1½"	21.69	22.13	22.57	23.01	23.45	0.29	0.23	142	150	158	166	174
	1¾"	23.55	24.15	24.75	25.35	25.95	0.31	0.25	168	179	190	201	212
10"–B	1¼"	21.13	21.45	21.77	22.09	22.41	0.26	0.22	190	195	200	205	210
	1½"	22.89	23.33	23.77	24.21	24.65	0.29	0.23	194	202	210	218	226
	1¾"	24.80	25.40	26.00	26.60	27.20	0.31	0.25	220	231	242	253	264
	2"	26.88	27.66	28.44	29.22	30.00	0.33	0.26	248	262	276	290	304
10"–C	2¼"	29.13	30.11	31.09	32.07	33.05	0.34	0.29	280	297	314	331	348
	2½"	32.36	33.58	34.80	36.02	37.24	0.36	0.31	329	350	371	392	413
12"–B	1½"	24.34	24.78	25.22	25.66	26.10	0.29	0.23	246	254	262	270	278
	1¾"	26.31	26.91	27.51	28.11	28.71	0.31	0.25	272	283	294	305	316
	2"	28.65	29.43	30.21	30.99	31.77	0.33	0.26	312	326	340	354	368
12"–C	2¼"	31.06	32.04	33.02	34.00	34.98	0.34	0.29	352	369	386	403	420
	2½"	34.34	35.56	36.78	38.00	39.22	0.36	0.31	401	422	443	464	485
12"–D	2¾"	37.26	38.72	40.18	41.64	43.10	0.38	0.33	443	469	495	521	547
14"–B	1½"	26.06	26.50	26.94	27.38	27.82	0.29	0.23	302	310	318	326	334
	1¾"	27.97	28.57	29.17	29.77	30.37	0.31	0.25	328	339	350	361	372
14"–C	2"	30.32	31.10	31.88	32.66	33.44	0.33	0.26	368	382	396	410	424
	2¼"	32.41	33.39	34.37	35.35	36.33	0.34	0.29	392	409	426	443	460
14"–D	2½"	36.21	37.43	38.65	39.87	41.09	0.36	0.31	465	486	507	528	549
14"–E	2¾"	39.24	40.70	42.16	43.62	45.08	0.38	0.33	515	541	567	593	619
14"–F	3"	40.66	42.40	44.14	45.88	47.62	0.39	0.34	584	615	646	677	708
16"–C	1¾"	29.89	30.49	31.09	31.69	32.29	0.31	0.25	404	415	426	437	448
	2"	31.82	32.60	33.38	34.16	34.94	0.33	0.26	424	438	452	466	480
	2¼"	33.86	34.84	35.82	36.80	37.78	0.34	0.29	448	465	482	499	516
16"–D	2½"	38.24	39.46	40.68	41.90	43.12	0.36	0.31	545	566	587	608	629
16"–E	2¾"	41.42	42.88	44.34	45.80	47.26	0.38	0.33	595	621	647	673	699
16"–F	3"	42.74	44.48	46.22	47.96	49.70	0.39	0.34	664	695	726	757	788
18"–C	2"	33.75	34.53	35.31	36.09	36.87	0.33	0.26	484	498	512	526	540
	2¼"	35.84	36.82	37.80	38.78	39.76	0.34	0.29	508	525	542	559	576
18"–D	2½"	40.11	41.33	42.55	43.77	44.99	0.36	0.31	601	622	643	664	685
18"–E	2¾"	43.40	44.86	46.32	47.78	49.24	0.38	0.33	659	685	711	737	763
18"–F	3"	45.03	46.77	48.51	50.25	51.99	0.39	0.34	740	771	802	833	864

8" Diameter wheels in plate pattern only.
10" Diameter wheels will be furnished in plate pattern unless otherwise specified.
See page 20 for information on wheel weights, wheel treads, etc.
See page 29 for axle clips.

"CARD" PLAIN BEARING TRUCKS

OPEN HUB WHEELS—SQUARE AXLES—NO PEDESTALS

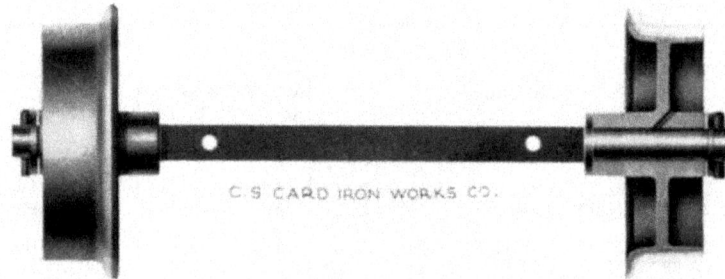

This truck is placed on the market to meet the demand for an inexpensive truck for short and slow haulage conditions.

It is regularly furnished with "CARD" semi-steel, heat-treated wheels with liberal bearing length and the rear of the hub counterbored for the axle washer.

The axles are a special analysis high carbon steel to resist wear.

The illustration shows a Style 6 axle with holes for bolting the truck to a car frame. It can be furnished with any style axle shown on page 28.

See page 15 for price list and weight of this truck.

Most mining men are familiar with this truck with its many variations with screw and bolted caps. We can furnish trucks with wheels with any of these features, and prices will be quoted on receipt of specifications.

However, where wheels with caps are desired, we recommend the trucks on the preceding pages, equipped with solid cap wheels.

"CARD" PLAIN BEARING TRUCKS
OPEN HUB WHEELS—SQUARE AXLES—NO PEDESTALS

Prices and weights listed below cover one truck. A truck consists of four wheels (spoke or plate pattern) and two axles, assembled and ready to attach to car.

When ordering or inquiring for prices, please specify the wheel diameter, the axle size and style number, and the track gauge. State whether spoke or plate pattern wheels are desired.

Lists cover a truck with Style No. 1 axles; other styles of axles carry extras. See page 28 for axle styles and numbers.

Wheel Diam. and Weight	Axle Size	PRICE PER TRUCK					Add for Style #2 Axles	Add for Style #6 Axles	WEIGHT (Lbs.) PER TRUCK				
		Track Gauge							Track Gauge				
		18"	24"	30"	36"	42"			18"	24"	30"	36"	42"
6"-C	1¼"	$15.62	$15.94	$16.26	$16.58	$16.90	$0.26	$0.22	98	103	108	113	118
	1½"	17.44	17.88	18.32	18.76	19.20	0.29	0.23	106	114	122	130	138
8"-C	1¼"	16.03	16.35	16.67	16.99	17.31	0.26	0.22	106	111	116	121	126
	1½"	18.22	18.66	19.10	19.54	19.98	0.29	0.23	130	138	146	154	162
	1¾"	20.07	20.67	21.27	21.87	22.47	0.31	0.25	156	167	178	189	200
10"-B	1¼"	17.18	17.50	17.82	18.14	18.46	0.26	0.22	158	163	168	173	178
	1½"	19.42	19.86	20.30	20.74	21.18	0.29	0.23	182	190	198	206	214
	1¾"	21.26	21.86	22.46	23.06	23.66	0.31	0.25	208	219	230	241	252
	2"	23.26	24.04	24.82	25.60	26.38	0.33	0.26	235	249	263	277	291
10"-C	2¼"	25.49	26.47	27.45	28.43	29.41	0.34	0.29	264	281	298	315	332
	2½"	28.49	29.71	30.93	32.15	33.37	0.36	0.31	306	327	348	369	390
12"-B	1½"	20.87	21.31	21.75	22.19	22.63	0.29	0.23	234	242	250	258	266
	1¾"	22.72	23.32	23.92	24.52	25.12	0.31	0.25	260	271	282	293	304
	2"	24.97	25.75	26.53	27.31	28.09	0.33	0.26	299	313	327	341	355
12"-C	2¼"	27.42	28.40	29.38	30.36	31.34	0.34	0.29	336	353	370	387	404
	2½"	30.41	31.63	32.85	34.07	35.29	0.36	0.31	378	399	420	441	462
12"-D	2¾"	33.27	34.73	36.19	37.65	39.11	0.38	0.33	422	448	474	500	526
14"-B	1½"	22.59	23.03	23.47	23.91	24.35	0.29	0.23	290	298	306	314	322
	1¾"	24.43	25.03	25.63	26.23	26.83	0.31	0.25	316	327	338	349	360
14"-C	2"	26.69	27.47	28.25	29.03	29.81	0.33	0.26	355	369	383	397	411
	2¼"	28.72	29.70	30.68	31.66	32.64	0.34	0.29	376	393	410	427	444
14"-D	2½"	32.29	33.51	34.73	35.95	37.17	0.36	0.31	442	463	484	505	526
14"-E	2¾"	35.30	36.76	38.22	39.68	41.14	0.38	0.33	494	520	546	572	598
14"-F	3"	36.68	38.42	40.16	41.90	43.64	0.39	0.34	562	593	624	655	686
16"-C	1¾"	26.36	26.96	27.56	28.16	28.76	0.31	0.25	392	403	414	425	436
	2"	28.15	28.93	29.71	30.49	31.27	0.33	0.26	411	425	439	453	467
	2¼"	30.22	31.20	32.18	33.16	34.14	0.34	0.29	432	449	466	483	500
16"-D	2½"	34.31	35.53	36.75	37.97	39.19	0.36	0.31	522	543	564	585	606
16"-E	2¾"	37.32	38.78	40.24	41.70	43.16	0.38	0.33	574	600	626	652	678
16"-F	3"	38.66	40.40	42.14	43.88	45.62	0.39	0.34	642	673	704	735	766
18"-C	2"	30.12	30.90	31.68	32.46	33.24	0.33	0.26	471	485	499	513	527
	2¼"	32.15	33.13	34.11	35.09	36.07	0.34	0.29	492	509	526	543	560
18"-D	2½"	36.13	37.35	38.57	39.79	41.01	0.36	0.31	578	599	620	641	662
18"-E	2¾"	39.35	40.81	42.27	43.73	45.19	0.38	0.33	638	664	690	716	742
18"-F	3"	41.00	42.74	44.48	46.22	47.96	0.39	0.34	718	749	780	811	842

6" and 8" Diameter wheels in plate pattern only.
10" Diameter wheels will be furnished in plate pattern unless otherwise specified.
See page 20 for information on wheel weights, wheel treads, etc.
See page 29 for axle clips.

"CARD" TIMKEN ROLLER BEARING SKIP TRUCKS

An easy-running skip truck, of a design that eliminates the many difficulties usually encountered in a skip truck. It combines "CARD" semi-steel heat-treated wheels and Timken roller bearings, both thoroughly time and mine tested.

The bearings are grease lubricated and protected against entrance of dirt.

Axles are a special analysis high carbon steel.

The standard truck specifications listed below are with square axles, which can attach direct to and support the skip bottom.

Any wheel diameter, axle size, or track gauge can be furnished.

Prices and weights listed cover a truck consisting of four wheels with caps, Timken roller bearings, and two high carbon steel axles, assembled and grease packed, ready for use.

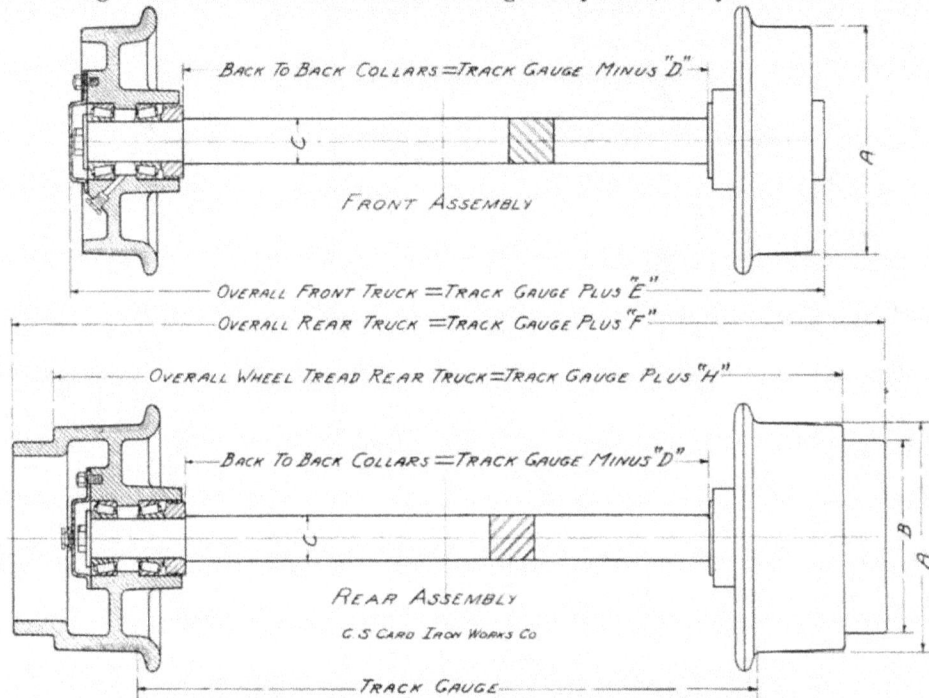

Specifications; Prices and Average Weights

A Wheel Diam. at Rail	B	C Axle Size *	D	E	F	H	PRICE PER TRUCK Track Gauge				WEIGHT (Lbs.) PER TRUCK Track Gauge			
							24"	30"	36"	42"	24"	30"	36"	42"
10"	8½"	1¾"	4⅛"	6⅛"	12⅜"	7⅝"	$61.58	$62.18	$62.78	$63.38	278	288	298	308
10"	8½"	2"	4⅜"	6⅜"	12⅜"	7⅝"	66.97	67.75	68.53	69.31	300	314	328	342
10"	8½"	2¼"	4⅜"	6⅜"	12⅜"	7⅝"	71.04	72.02	73.00	73.98	328	346	364	382
10"	8½"	2½"	4⅜"	5⅞"	12⅜"	7⅝"	75.96	77.18	78.40	79.62	363	385	407	429
12"	10½"	1¾"	4⅛"	6⅛"	12⅜"	7⅝"	63.79	64.39	64.99	65.59	368	378	388	398
12"	10½"	2"	4⅜"	6⅜"	12⅜"	7⅝"	69.19	69.97	70.75	71.53	390	404	418	432
12"	10½"	2¼"	4⅜"	6⅜"	12⅜"	7⅝"	73.29	74.27	75.25	76.23	418	436	454	472
12"	10½"	2½"	4⅜"	5⅞"	12⅜"	7⅝"	78.21	79.43	80.65	81.87	453	475	497	519

*—Square axle size. Finished diameter at bearings is ¼" less.

Rear wheels with a straight wide tread or with the skip ring tread a different diameter or width than the standard shown can be furnished at a slight extra.

Wheels with special flange 1¾" high can be furnished where conditions require at a slight extra. Standard flange heights are ⅞" for 10" diameter and 1" for 12" diameter wheels.

Skip trucks with round special analysis high carbon steel axles mounted in pedestals or boxes can be furnished on receipt of your requirements.

"CARD" SPECIAL TRUCKS

While our standard design trucks shown on preceding pages will cover most of the requirements for trucks in the mine car and industrial car field, they do not by any means cover our entire range of design and manufacture. Conditions arise which make it advisable or necessary to use other than our standard design. A few of these are shown.

If you have a truck problem, submit it to us. Our engineers will be glad to offer suggestions.

A spring mounted truck, with the spring pedestals mounted to the car frame inside the wheels. Roller bearings are mounted in the wheels.

A bolster truck with top bolster removed:—spring mounting:—wheels tight on the axle:—roller bearings mounted in outside journal boxes.

A bolster truck with top bolster removed:—spring mounting:—roller bearings are mounted in the wheels which turn on the axle.

A truck with outboard flat spring mounting and independent truck frame. Roller bearings are mounted in the wheels which turn on the axle. This design is adapted advantageously to narrow gauges and where room for installation is limited. When wide car bodies are used in relation to the track gauge, this truck increases the stability and reduces the swaying of the car.

"CARD" SEMI-STEEL HEAT-TREATED WHEELS

A wheel floor showing wheel flasks. An overhead system and hoists facilitate the handling of all equipment and pouring of the hot metal.

A corner in CARD'S metallurgical laboratory.

Un retouched Photo

Broken test bars, showing varying depth of chill and structure of metal. No CARD wheel is ever poured until the test bar shows satisfactory metal for the purpose desired.

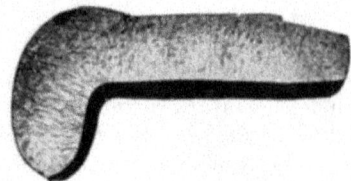

An unretouched photo of a rim section from a "CARD" Semi-Steel Heat-Treated Wheel, showing depth and uniformity of chill.

"CARD" SEMI-STEEL HEAT-TREATED WHEELS

We have specialized in the manufacture of mine car wheels since 1892. With our modern well equipped wheel foundry, we are in a position to furnish the trade with a strictly high grade wheel made to meet their requirements. We will furnish any wheel not covered by patent.

From our long experience and with our modern equipment, we are in a position to control the many factors entering into the production of mine car wheels. Aside from design and the proper relationship of rim, spokes or plate, and hub there are many other factors to control. We mention but a few such as proper metal mixture for each type and weight of wheel, speed and method of pouring, temperature of metal, and proper annealing.

All of our wheel mixtures carry a high percentage of steel, making a semi or part steel casting. When properly handled this mixture makes a wheel with close-grained, tough metal that withstands lots of abuse, and with a high-class chilled wheel tread that withstands abrasion and wear.

All our chilled semi-steel wheels are heat-treated to develop maximum strength.

When ordering or asking for prices on other than our standard design wheels, please furnish us a drawing or send us a sample wheel—not too badly worn.

For your convenience the illustration on page 20 gives the names commonly applied to the different wheel parts.

"CARD" SEMI-STEEL MOLYCHRO HEAT-TREATED WHEELS

We pioneered in the use of alloys in wheels and in 1928 introduced our MOLYCHRO wheel for those mines that have severe haulage or desire a better wheel.

Molybdenum and chrome are added to our time-tested semi-steel wheel metal as a base. The addition of these alloying metals has a distinct and pronounced effect on the structure of the base metal, giving it improved physical properties, with resultant added service.

Field reports and repeat orders indicate that the increased cost of the MOLYCHRO alloy wheel is justified.

It can be furnished to any unpatented wheel design.

We are continuously working with the different alloying metals to improve our alloy wheels. As these are developed and as experience dictates we will furnish these wheels to the trade.

"CARD" SEMI-STEEL
HEAT-TREATED WHEELS

STANDARD WHEEL DATA

It is only natural that in the wide field to which we supply wheels, conditions will vary greatly. We make wheels to a number of different weights and widths of tread.

Haulage factors such as the weight of the load, speed and length of haul, track conditions, etc., all enter into the selection of the wheel weight and axle size that should be used to obtain the most economical and satisfactory service.

In our lists you will note the use of letters following the wheel diameters. These letters designate the weight of the wheel.

The following table shows our standard practice and based on our long experience will cover the large majority of haulage conditions.

The table designates the weight of the wheel and the width of the wheel tread for each wheel diameter and the size of axle with which the wheel is used. On all orders, wheels will be furnished according to this table, unless otherwise specified.

STANDARD WHEEL DATA

Wheel Diameter	Rim and Spoke or Plate Weight	Standard Tread Width	Axle Size	Wheel Diameter	Rim and Spoke or Plate Weight	Standard Tread Width	Axle Size
6"	C	2¼"	1¼"	14"	B	3"	1½"
6"	C	2¼"	1½"	14"	B	3"	1¾"
				14"	C	3"	2"
8"	C	2¼"	1¼"	14"	C	3"	2¼"
8"	C	2¼"	1½"	14"	D	3"	2½"
8"	C	2¼"	1¾"	14"	E	3¼"	2¾"
				14"	F	3¼"	3"
10"	B	2½"	1¼"				
10"	B	2½"	1½"	16"	C	3"	1¾"
10"	B	2½"	1¾"	16"	C	3"	2"
10"	B	2½"	2"	16"	C	3"	2¼"
10"	C	2½"	2¼"	16"	D	3¼"	2½"
10"	C	2½"	2½"	16"	E	3¼"	2¾"
				16"	F	3¼"	3"
12"	B	2¾"	1½"				
12"	B	2¾"	1¾"	18"	C	3¼"	2"
12"	B	2¾"	2"	18"	C	3¼"	2¼"
12"	C	3"	2¼"	18"	D	3½"	2½"
12"	C	3"	2½"	18"	E	3½"	2¾"
12"	D	3"	2¾"	18"	F	3½"	3"

6" and 8" diameter standard wheels furnished plate pattern only.

10" diameter wheels furnished plate pattern unless otherwise specified. Spoke pattern wheels can be furnished.

12", 14", 16", and 18" diameter wheels furnished either spoke or plate pattern as is specified.

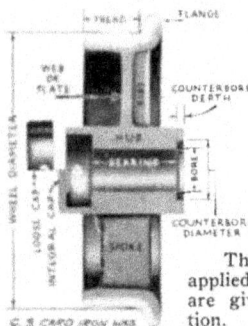

Wheel Tread Extras

For wheels with treads wider than standard, add to each wheel for each ¼ inch of additional tread width, the following extras.

The names commonly applied to parts of a wheel are given in the illustration.

Wheel Diam.	Price Extra	Weight Extra, Pounds
6"	$0.02	¾
8"	0.03	1
10"	0.03	1¼
12"	0.04	1½
14"	0.06	2½
16"	0.08	3
18"	0.09	3½

"CARD" ROLLER BEARING WHEELS WITH SOLID CAP

SPOKE PATTERN

This list covers our standard spoke pattern semi-steel heat-treated solid cap roller bearing wheels as used on the trucks on pages 4 and 8.

Repair wheels will be furnished with the renewable hard steel split bushings unless otherwise specified by you.

See page 22 for solid cap roller bearing wheels of plate pattern.

For standard wheel data see page 20.

Specifications: Prices and Average Weights in Pounds, Each Finished Wheel (without plugs)

Wheel Diam. and Weight	Axle Size	WHEEL ONLY Not Bushed		WHEEL ONLY With Bushing		Wheel Diam and Weight	Axle Size	WHEEL ONLY Not Bushed		WHEEL ONLY With Bushing	
		Price	Weight	Price	Weight			Price	Weight	Price	Weight
10″-B	1½″	$3.38	42	$3.85	43	14″-E	2½″	$5.54	99	$6.78	102
	1¾″	3.53	44	4.09	45		2¾″	5.80	104	7.20	107
	2″	3.77	49	4.62	51	14″-F	2¾″	5.99	111	7.39	114
10″-C	2″	3.85	52	4.70	54		3″	6.25	115	8.30	119
	2¼″	4.03	54	5.09	56	16″-C	1¾″	5.10	92	5.66	93
	2½″	4.24	58	5.48	61		2″	5.32	97	6.17	99
12″-B	1½″	3.84	56	4.31	57		2¼″	5.47	98	6.53	100
	1¾″	3.99	58	4.55	59	16″-D	2¼″	5.63	103	6.69	105
	2″	4.24	63	5.09	65		2½″	5.80	106	7.04	109
12″-C	2″	4.36	68	5.21	70	16″-E	2½″	6.10	117	7.34	120
	2¼″	4.52	69	5.58	71		2¾″	6.34	122	7.74	125
	2½″	4.72	73	5.96	76	16″-F	2¾″	6.45	126	7.85	129
12″-D	2½″	4.82	77	6.06	80		3″	6.70	129	8.75	133
	2¾″	5.08	82	6.48	85	18″-C	2″	5.98	114	6.83	116
14″-B	1½″	4.26	68	4.73	69		2¼″	6.11	115	7.17	117
	1¾″	4.44	70	5.00	71	18″-D	2¼″	6.35	124	7.41	126
14″-C	1¾″	4.72	80	5.28	81		2½″	6.54	127	7.78	130
	2″	4.85	81	5.70	83	18″-E	2½″	6.84	138	8.08	141
	2¼″	5.01	82	6.07	84		2¾″	7.10	143	8.50	146
14″-D	2¼″	5.10	86	6.16	88	18″-F	2¾″	7.37	153	8.77	156
	2½″	5.31	90	6.55	93		3″	7.57	155	9.62	159

Standard Wheels are in Bold Face Type.

For standard width wheel treads and price extras for special width treads, see page 20.
Prices on application for these wheels made of our MOLYCHRO wheel metal. See page 19.

"CARD" ROLLER BEARING WHEELS WITH SOLID CAP

PLATE PATTERN

This list covers our standard plate pattern semi-steel heat-treated solid cap roller bearing wheels as used on the trucks on pages 4 and 8.

Repair wheels will be furnished with renewable hard steel split bushings unless otherwise specified by you.

See page 21 for solid cap roller bearing wheels of spoke pattern.

For standard wheel data see page 20.

Specifications; Prices and Average Weights in Pounds, Each Finished Wheel (without plugs)

Wheel Diam. And Weight	Axle Size	WHEEL ONLY Not Bushed		WHEEL ONLY With Bushing		Wheel Diam. And Weight	Axle Size	WHEEL ONLY Not Bushed		WHEEL ONLY With Bushing	
		Price	Weight	Price	Weight			Price	Weight	Price	Weight
8"–C	1½"	$3.11	33	$3.58	34	14"–E	2½"	$5.80	106	$7.04	109
	1¾"	3.27	35	3.83	36		2¾"	6.08	112	7.48	115
10"–B	1½"	3.38	42	3.85	43	14"–F	2¾"	6.24	118	7.64	121
	1¾"	3.53	44	4.09	45		3"	6.53	123	8.58	127
	2"	3.77	49	4.62	51	16"–C	1¾"	5.44	102	6.00	103
10"–C	2"	3.85	52	4.70	54		2"	5.59	104	6.44	106
	2¼"	4.03	54	5.09	56		2¼"	5.73	105	6.79	107
	2½"	4.18	56	5.42	59	16"–D	2¼"	5.87	110	6.93	112
12"–B	1½"	3.95	60	4.42	61		2½"	6.07	114	7.31	117
	1¾"	4.12	62	4.68	63	16"–E	2½"	6.42	126	7.66	129
	2"	4.32	66	5.17	68		2¾"	6.65	131	8.05	134
12"–C	2"	4.42	70	5.27	72	16"–F	2¾"	6.84	138	8.24	141
	2¼"	4.61	72	5.67	74		3"	7.08	141	9.13	145
	2½"	4.91	80	5.15	83	18"–C	2"	6.35	126	7.20	128
12"–D	2½"	5.06	85	6.30	88		2¼"	6.51	127	6.57	129
	2¾"	5.31	90	6.71	93	18"–D	2¼"	6.64	132	7.70	134
14"–B	1½"	4.55	76	5.02	77		2½"	6.84	136	8.08	139
	1¾"	4.72	78	5.28	79	18"–E	2½"	7.20	149	8.44	152
14"–C	1¾"	4.90	85	5.46	86		2¾"	7.43	153	8.83	156
	2"	5.06	87	5.91	89	18"–F	2¾"	7.72	163	9.12	166
	2¼"	5.23	89	6.29	91		3"	7.95	167	10.00	171
14"–D	2¼"	5.35	93	6.41	95						
	2½"	5.53	96	6.77	99						

Standard Wheels are in Bold Face Type.

For standard width wheel treads and price extras for special width treads, see page 20.
Prices on application for these wheels made of our MOLYCHRO wheel metal. See page 19.

"CARD" TIMKEN ROLLER BEARING WHEELS

SPOKE AND PLATE PATTERNS

This list covers our standard semi-steel heat-treated wheels for use with Timken roller bearing truck shown on page 6.

Prices and weights shown cover the finished wheel only; without caps, cups, cap bolts or washers, or grease fittings.

For standard wheel data see page 20.

Specifications; Prices and Average Weights in Pounds, Each Finished Wheel

Wheel Diam. And Weight	Axle Size	SPOKE WHEEL Wheel Only Price	SPOKE WHEEL Wheel Only Weight	PLATE WHEEL Wheel Only Price	PLATE WHEEL Wheel Only Weight	Wheel Diam. And Weight	Axle Size	SPOKE WHEEL Wheel Only Price	SPOKE WHEEL Wheel Only Weight	PLATE WHEEL Wheel Only Price	PLATE WHEEL Wheel Only Weight
8"–C	1¾"	....	....	$3.88	37	14"–E	2½"	$5.88	90	$6.08	95
							2¾"	6.08	93	6.26	98
10"–B	1¾"	$4.04	43	4.04	43	14"–F	2¾"	6.25	99	6.44	104
	2"	4.21	45	4.21	45		3"	6.61	104	6.74	107
10"–C	2"	4.26	48	4.26	48	16"–C	2"	5.68	90	5.83	93
	2¼"	4.51	51	4.51	51		2¼"	5.87	92	5.99	95
	2½"	4.73	55	4.73	55	16"–D	2¼"	5.98	96	6.14	100
12"–B	1¾"	4.55	59	4.66	63		2½"	6.15	98	6.33	102
	2"	4.62	60	4.72	64	16"–E	2½"	6.42	107	6.65	114
12"–C	2"	4.78	63	4.88	67		2¾"	6.57	109	6.82	116
	2¼"	4.91	64	5.03	68	16"–F	2¾"	6.76	116	7.01	123
	2½"	5.08	65	5.26	71		3"	7.10	120	7.31	126
12"–D	2½"	5.26	71	5.38	76	18"–C	2"	6.29	106	6.64	117
	2¾"	5.42	74	5.54	78		2¼"	6.46	108	6.80	118
14"–B	1¾"	4.95	69	4.97	68	18"–D	2¼"	6.57	112	6.97	124
14"–C	1¾"	5.00	71	5.22	77		2½"	6.76	114	7.16	126
	2"	5.17	73	5.38	79	18"–E	2½"	7.07	125	7.49	139
	2¼"	5.35	75	5.54	81		2¾"	7.21	127	7.64	140
14"–D	2¼"	5.42	78	5.65	85	18"–F	2¾"	7.54	138	7.93	150
	2½"	5.64	81	5.80	85		3"	7.83	141	8.20	152

Standard Wheels are in Bold Face Type.

For standard width wheel treads and price extras for special width treads, see page 20.
Prices on application for these wheels made of our MOLYCHRO wheel metal. See page 19.

WHEEL CAPS (for above wheels)

Specifications; Prices and Average Weights in Pounds

Axle Size		1¾"	2"	2¼"	2½"	2¾"	3"
Cap Only	Price, each	$0.66	$0.66	$0.69	$0.78	$0.78	$0.84
	Weight, each	2	2	3	4	4	4
Cap with bolts and lock washers	Price, each	$0.76	$0.76	$0.79	$0.97	$0.97	$1.03
	Weight, each	2	2	3	5	5	5

"CARD" PLAIN BEARING WHEELS WITH SOLID CAP

SPOKE AND PLATE PATTERNS

This list covers our standard semi-steel heat-treated solid cap wheels with oil ports as used on the trucks on pages 10 and 12.

For standard wheel data see page 20.

Specifications; Prices and Average Weights in Pounds, Each Finished Wheel (without plugs)

Wheel Diam. And Weight	Axle Size	SPOKE WHEEL Wheel Only Price	SPOKE WHEEL Wheel Only Weight	PLATE WHEEL Wheel Only Price	PLATE WHEEL Wheel Only Weight
8"–C	1½"	....	....	$3.95	28
	1¾"	....	....	4.14	32
10"–B	1¼"	$4.23	42	4.23	42
	1½"	4.31	43	4.31	43
	1¾"	4.51	45	4.51	45
	2"	4.70	48	4.70	48
10"–C	2"	4.76	50	4.76	50
	2¼"	4.90	52	4.90	52
	2½"	5.23	60	5.23	60
12"–B	1½"	4.73	54	4.98	63
	1¾"	4.95	58	5.18	67
	2"	5.22	64	5.28	67
12"–C	2"	5.28	67	5.35	68
	2¼"	5.47	70	5.57	72
	2½"	5.83	77	6.01	84
12"–D	2½"	5.95	82	6.17	90
	2¾"	6.11	84	6.34	92
14"–B	1½"	5.23	68	5.44	74
	1¾"	5.44	72	5.64	78
14"–C	1¾"	5.53	76	5.74	82
	2"	5.72	78	5.93	84
	2¼"	5.87	80	6.20	90
14"–D	2¼"	6.07	87	6.38	97
	2½"	6.38	93	6.57	100

Wheel Diam. And Weight	Axle Size	SPOKE WHEEL Wheel Only Price	SPOKE WHEEL Wheel Only Weight	PLATE WHEEL Wheel Only Price	PLATE WHEEL Wheel Only Weight
14"–E	2½"	$6.54	100	$6.79	106
	2¾"	6.70	102	6.97	109
14"–F	2¾"	6.90	108	7.12	115
	3"	7.03	110	7.31	118
16"–C	1¾"	6.01	91	6.25	98
	2"	6.15	92	6.40	99
	2¼"	6.30	94	6.54	101
16"–D	2¼"	6.76	110	7.03	118
	2½"	6.97	113	7.26	121
16"–E	2½"	7.17	120	7.47	129
	2¾"	7.34	122	7.74	135
16"–F	2¾"	7.49	128	7.90	140
	3"	7.64	130	8.03	142
18"–C	2"	6.72	107	7.11	119
	2¼"	6.88	109	7.27	121
18"–D	2¼"	7.37	126	7.67	135
	2½"	7.53	127	7.82	136
18"–E	2½"	7.76	136	8.13	147
	2¾"	7.93	138	8.25	148
18"–F	2¾"	8.19	147	8.51	157
	3"	8.33	149	8.72	161

Standard Wheels are in Bold Face Type.

For standard width wheel treads and price extras for special width treads, see page 20.
Prices on application for these wheels made of our MOLYCHRO wheel metal. See page 19.

"CARD" PLAIN BEARING WHEELS WITH OPEN HUB

SPOKE AND PLATE PATTERNS

This list covers our standard semi-steel heat-treated open hub plain bearing wheels as used on the truck on page 14.

For standard wheel data, see page 20.

OPEN HUB
WHEEL

Specifications; Prices and Average Weights in Pounds, Each Finished Wheel

Wheel Diam. And Weight	Axle Size	SPOKE WHEEL Wheel Only		PLATE WHEEL Wheel Only		Wheel Diam. And Weight	Axle Size	SPOKE WHEEL Wheel Only		PLATE WHEEL Wheel Only	
		Price	Weight	Price	Weight			Price	Weight	Price	Weight
6″–C	1¼″	. . . .	. . . . .	$2.71	19	14″–D	2¼″	$5.07	83	$5.41	93
	1½″	. . . .	. . . . .	2.80	19		2½″	5.32	88	5.54	95
8″–C	1¼″	. . . .	. . . . .	2.83	21	14″–E	2½″	5.53	95	5.78	102
	1½″	. . . .	. . . . .	3.02	25		2¾″	5.65	97	5.92	105
	1¾″	. . . .	. . . . .	3.21	29	14″–F	2¾″	5.83	103	6.07	110
10″–B	1¼″	$3.17	34	3.17	34		3″	5.99	105	6.25	113
	1½″	3.38	38	3.38	38	16″–C	1¾″	5.07	88	5.32	95
	1¾″	3.57	42	3.57	42		2″	5.19	89	5.46	96
	2″	3.75	45	3.75	45		2¼″	5.32	90	5.57	97
10″–C	2″	3.79	47	3.79	47	16″–D	2¼″	5.79	106	6.01	112
	2¼″	3.94	48	3.94	48		2½″	5.93	108	6.21	116
	2½″	4.21	54	4.21	55	16″–E	2½″	6.11	115	6.44	124
12″–B	1½″	3.80	51	4.07	60		2¾″	6.25	117	6.67	130
	1¾″	3.99	55	4.25	64	16″–F	2¾″	6.42	123	6.80	135
	2″	4.25	61	4.35	64		3″	6.57	125	6.97	137
12″–C	2″	4.35	64	4.41	65	18″–C	2″	5.78	104	6.15	116
	2¼″	4.51	66	4.57	68		2¼″	5.89	105	6.29	117
	2½″	4.78	72	4.97	79	18″–D	2¼″	6.34	120	6.55	126
12″–D	2½″	4.90	77	5.12	85		2½″	6.46	122	6.79	131
	2¾″	5.06	79	5.27	87	18″–E	2½″	6.72	131	7.08	142
14″–B	1½″	4.32	65	4.54	71		2¾″	6.86	133	7.18	143
	1¾″	4.51	69	4.72	75	18″–F	2¾″	7.11	142	7.46	152
14″–C	1¾″	4.64	74	4.85	80		3″	7.27	144	7.64	156
	2″	4.76	75	4.98	81						
	2¼″	4.88	76	5.22	86						

Standard Wheels are in Bold Face Type.

For standard width wheel treads and price extras for special width treads, see page 20.
Prices on application for these wheels made of our MOLYCHRO wheel metal. See page 19.

"CARD" STANDARD WHEEL PARTS

This illustration shows the few parts used in "CARD" standard solid cap roller bearing wheels. Plain bearing wheels have even fewer parts.

Prices listed are for standard equipment.

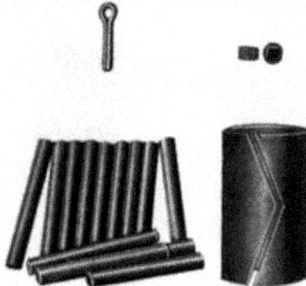

Renewable Hard Steel Split Bushings

These are made from steel made to our special analysis and not from the ordinary cold rolled strip steel on the market.

	Axle Size						
	1½"	1¾"	2"	2¼"	2½"	2¾"	3"
Price, each.......	$0.47	$0.56	$0.85	$1.06	$1.24	$1.40	$2.05
Weight, (Lbs.) each	0.78	0.97	1.67	2.04	2.49	2.85	4.31

"Card" Solid Roller Bearings

A special grade of steel, which we have thoroughly tested, is used in these bearings.

	Axle Size						
	1½"	1¾"	2"	2¼"	2½"	2¾"	3"
Price, Per Set (for one wheel)....	$0.60	$0.78	$0.96	$1.16	$1.50	$1.69	$2.37
Number of rollers, per set.......	12	12	14	14	13	11	12
Weight (Lbs.) per set............	3.2	4.4	5.7	7.5	11.0	12.7	19.4

Countersunk Wheel Plugs

	Size		
	½"	¾"	1"
Price, per 100 plugs	$4.00	$6.00	$8.00
Weight, (Lbs.) per 100 plugs	5.5	10.0	20.0

Axle Cotters—Special Alloy Steel

We make these cotters from a special analysis chrome alloy steel of our own specifications. Tests show that they have approximately thirty-five times the life of the ordinary soft steel cotter.

We can furnish these in any length, in ½" or ⅝" cotters. The cut shows clearly that the length of a cotter is specified and measured under the head.

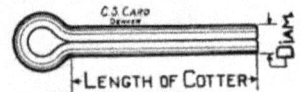

Prices and Weights in Pounds per 100 Cotters:

½" Cotter	Length Under Head								
	2"	*2¼"	*2½"	2¾"	*3"	*3¼"	3½"	4"	5"
	$8.83	$9.10	$9.32	$9.55	$9.77	$10.02	$10.27	$10.73	$11.69
	14.4	15.7	16.9	18.1	19.3	20.5	21.8	24.3	29.2

⅝" Cotter	Length Under Head								
	3¼"	*3½"	3¾"	*4"	*4¼"	4½"	5"		
	$13.70	$14.10	$14.48	$14.85	$15.24	$15.62	$16.38		
	34.4	36.4	38.4	40.3	42.3	44.3	48.2		

*Standard sizes carried in stock.

"CARD" ROUND MINE CAR AXLES

The round axle illustrated is for use with pedestals or journal boxes and of the type used on the trucks on pages 4 and 10.

All our round axles are made from special analysis high carbon steel to resist wear and bending.

We can furnish any style of round axle on receipt of a dimensioned sketch from you.

Prices listed cover axles of the style illustrated.

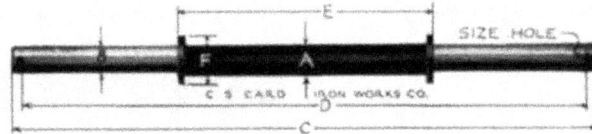

Specifications; Prices and Average Weights in Pounds, Each

Axle Size		1½″	1¾″	2″	2¼″	2½″	2¾″	3″
18″ Track Gauge	Price	$2.47	$2.99	$3.17	$4.09	$4.88	$5.67	$6.38
	Weight	13	17	24	30	38	48	55
24″ Track Gauge	Price	$2.66	$3.25	$3.51	$4.53	$5.42	$6.31	$7.16
	Weight	16	21	29	37	46	58	67
30″ Track Gauge	Price	$2.85	$3.51	$3.85	$4.97	$5.96	$6.95	$7.94
	Weight	19	25	34	44	54	68	79
36″ Track Gauge	Price	$3.04	$3.77	$4.19	$5.41	$6.50	$7.59	$8.72
	Weight	22	29	39	51	62	78	91
42″ Track Gauge	Price	$3.23	$4.03	$4.53	$5.85	$7.04	$8.23	$9.50
	Weight	25	33	44	58	70	88	103

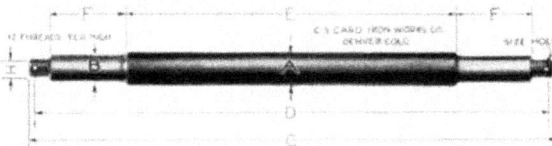

A high carbon steel axle for use with CARD-Timken roller bearing trucks as illustrated on page 6 is shown without dust collars. The pedestals or boxes are assembled on the axle before the dust collars are shrunk on. For this reason we will furnish the bare axle without the dust collars unless the collars are ordered to be either furnished shrunk on or shipped separately.

The following list covers the bare axle only.

Specifications; Prices and Average Weights in Pounds, Each, Without Dust Collars

Axle Size		1¾″	2″	2¼″	2½″	2¾″	3″
18″ Track Gauge	Price	$2.76	$3.45	$4.04	$4.71	$5.54	$6.32
	Weight	15	21	25	33	39	45
24″ Track Gauge	Price	$3.02	$3.79	$4.48	$5.25	$6.18	$7.10
	Weight	19	26	32	41	49	57
30″ Track Gauge	Price	$3.28	$4.13	$4.92	$5.79	$6.82	$7.88
	Weight	23	31	39	49	59	69
36″ Track Gauge	Price	$3.54	$4.47	$5.36	$6.33	$7.46	$8.66
	Weight	27	36	46	57	69	81
42″ Track Gauge	Price	$3.80	$4.81	$5.80	$6.87	$8.10	$9.44
	Weight	31	41	53	65	79	93

"CARD" SQUARE MINE CAR AXLES

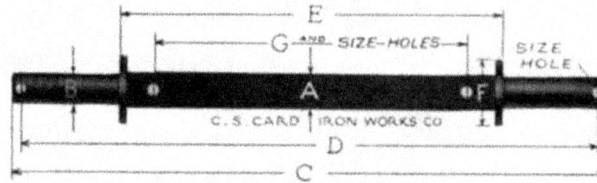

The square axle illustrated is the type used on the trucks shown on pages 12 and 14. A style No. 6 axle is used in the illustration but any one of the six styles shown below can be furnished. Styles No. 1, No. 2, and No. 6 are most often used.

Square axles of any design can be furnished on receipt of a drawing or dimensioned sketch from you. The illustration shows the principal dimensions required.

All our square axles are made from special analysis high carbon steel to resist wear and bending.

Prices on styles of square axles not listed will be furnished on receipt of specifications.

Specifications; Prices and Average Weights in Pounds, Each

AXLE SIZE		1¼″	1½″	1¾″	2″	2¼″	2½″	2¾″	3″
18″ Track Gauge	Axle No. 1, Price	$3.22	$3.82	$4.47	$5.15	$5.98	$6.85	$7.85	$9.38
	Axle No. 6, Price	3.47	4.08	4.75	5.44	6.31	7.20	8.22	9.76
	Axle No. 2, Price	3.51	4.15	4.82	5.52	6.36	7.26	8.28	9.82
	Weight	9	14	20	26	32	41	50	69
24″ Track Gauge	Axle No. 1, Price	$3.40	$4.07	$4.81	$5.59	$6.53	$7.54	$8.67	$10.36
	Axle No. 6, Price	3.65	4.33	5.09	5.88	6.86	7.89	9.04	10.74
	Axle No. 2, Price	3.69	4.40	5.16	5.96	6.91	7.95	9.10	10.80
	Weight	12	18	25	33	41	52	63	84
30″ Track Gauge	Axle No. 1, Price	$3.58	$4.32	$5.15	$6.03	$7.08	$8.23	$9.49	$11.34
	Axle No. 6, Price	3.83	4.58	5.43	6.32	7.41	8.58	9.86	11.72
	Axle No. 2, Price	3.87	4.65	5.50	6.40	7.46	8.64	9.92	11.78
	Weight	15	22	30	40	50	63	76	99
36″ Track Gauge	Axle No. 1, Price	$3.76	$4.57	$5.49	$6.47	$7.63	$8.92	$10.31	$12.32
	Axle No. 6, Price	4.01	4.83	5.77	6.76	7.96	9.27	10.68	12.70
	Axle No. 2, Price	4.05	4.90	5.84	6.84	8.01	9.33	10.74	12.76
	Weight	18	26	35	47	59	74	89	114
42″ Track Gauge	Axle No. 1, Price	$3.94	$4.82	$5.83	$6.91	$8.18	$9.61	$11.13	$13.30
	Axle No. 6, Price	4.19	5.08	6.11	7.20	8.51	9.96	11.50	13.68
	Axle No. 2, Price	4.23	5.15	6.18	7.28	8.56	10.02	11.56	13.74
	Weight	21	30	40	54	68	85	102	129

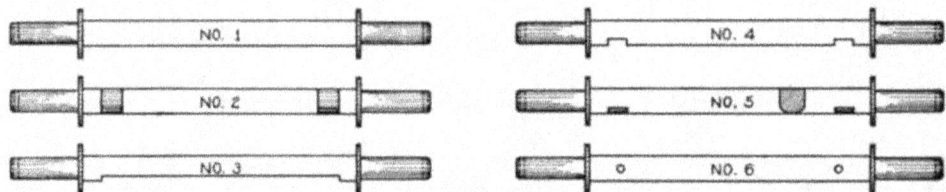

NO. 1

NO. 4

NO. 2

NO. 5

NO. 3

NO. 6

THE C.S. CARD IRON WORKS CO.

"CARD" AXLE CLIPS

Illustrated are standard axle clips which we use in attaching pedestals or axles to car bottoms or frames. The illustration on page 30 shows their use very clearly on wood bottom construction.

Where practical we make them continuous from axle to axle, helping to maintain wheel base. Also used at times with one short clip to a pedestal or axle and for attaching to structural frames, requiring four short clips to a truck.

They are formed to fit either a round seat or pedestal (as shown) or the square of an axle.

Can be made to meet most any requirement.

Unless you specify a preference we will furnish clips of the proper size and weight.

Specifications; Prices and Average Weights in Pounds, per Pair

Bar Size	WHEEL BASE							
	18″	20″	24″	28″	30″	36″	42″	48″
2″ x ½″	$3.10	$3.14	$3.28	$3.40	$3.46	$3.64	$3.82	$4.00
	23.8	25.0	27.2	29.4	30.6	34.0	37.4	40.8
2½″ x ½″	$3.40	$3.48	$3.62	$3.78	$3.88	$4.08	$4.28	$4.48
	29.8	31.2	34.0	36.8	38.2	42.4	46.6	50.8
3″ x ½″	$3.72	$3.80	$4.00	$4.18	$4.26	$4.54	$4.82	$5.10
	35.8	37.4	40.8	44.2	45.8	51.0	56.2	61.4
2½″ x ⅝″	$3.80	$3.90	$4.08	$4.26	$4.36	$4.64	$4.92	$5.20
	37.2	39.0	42.6	46.0	47.8	53.2	58.6	64.0
3″ x ⅝″	$4.22	$4.32	$4.56	$4.78	$4.88	$5.22	$5.56	$5.90
	44.6	46.8	51.0	55.2	57.4	63.8	70.2	76.6

"CARD" STANDARD TRUCK CROSS CHANNELS

For wood bottom cars, as used more particularly in coal mine car construction, we recommend the use of these channels for heavy loads and substantial construction. They reinforce and stiffen the bottom and make a better bearing for the pedestals. See page 30 for this construction.

Standard Style "A" pedestals are regularly used with these channels.

Specifications; Prices and Average Weights in Pounds, per Pair

³⁄₁₆″ Pressed Steel		Track Gauge			
		30″	36″	40″	42″
7″ Channel for Pedestals of 2″ to 2¾″ Axle Size	Price	$2.66	$3.00	$3.22	$3.32
	Weight	19.8	26.2	30.4	32.6
8″ Channel for Pedestals of 3″ Axle Size	Price	$2.76	$3.12	$3.38	$3.48
	Weight	21.4	28.4	33.2	35.4

"CARD" STANDARD PEDESTALS

The upper illustration shows our four standard styles of pedestals, with bells for the braided flax packing, as used with our "CARD" plain and roller bearing trucks with round axles on pages 4 and 10.

The lower illustration shows our four standard styles of plain pedestals as used with CARD-Timken roller bearing trucks with round axles on page 6.

Applications of these standard pedestals and of special pedestals are shown on page 31.

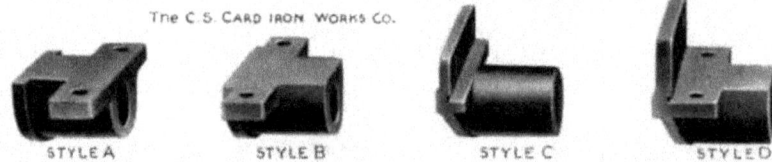

BOTTOM CONSTRUCTION

Shown is the method usually used in attaching our round axle trucks to wood car bottoms. Style "A" pedestals are used either with or without steel cross channels. Channels give a substantial bearing across the car bottom, making a very strong construction. We recommend this construction for heavy duty.

Each pedestal is fastened by two bolts and in addition is held in true alignment by the axle clips, which are formed to fit the pedestals and according to the wheel base. Note that there are twenty bolts used in this construction to hold the truck to the car bottom.

It can be readily seen from this construction, which has been thoroughly tested, that there is no way for the truck to be thrown out of alignment.

We recommend the axle clip construction. The clip strengthens and protects the pedestal against breakage.

PEDESTAL APPLICATIONS

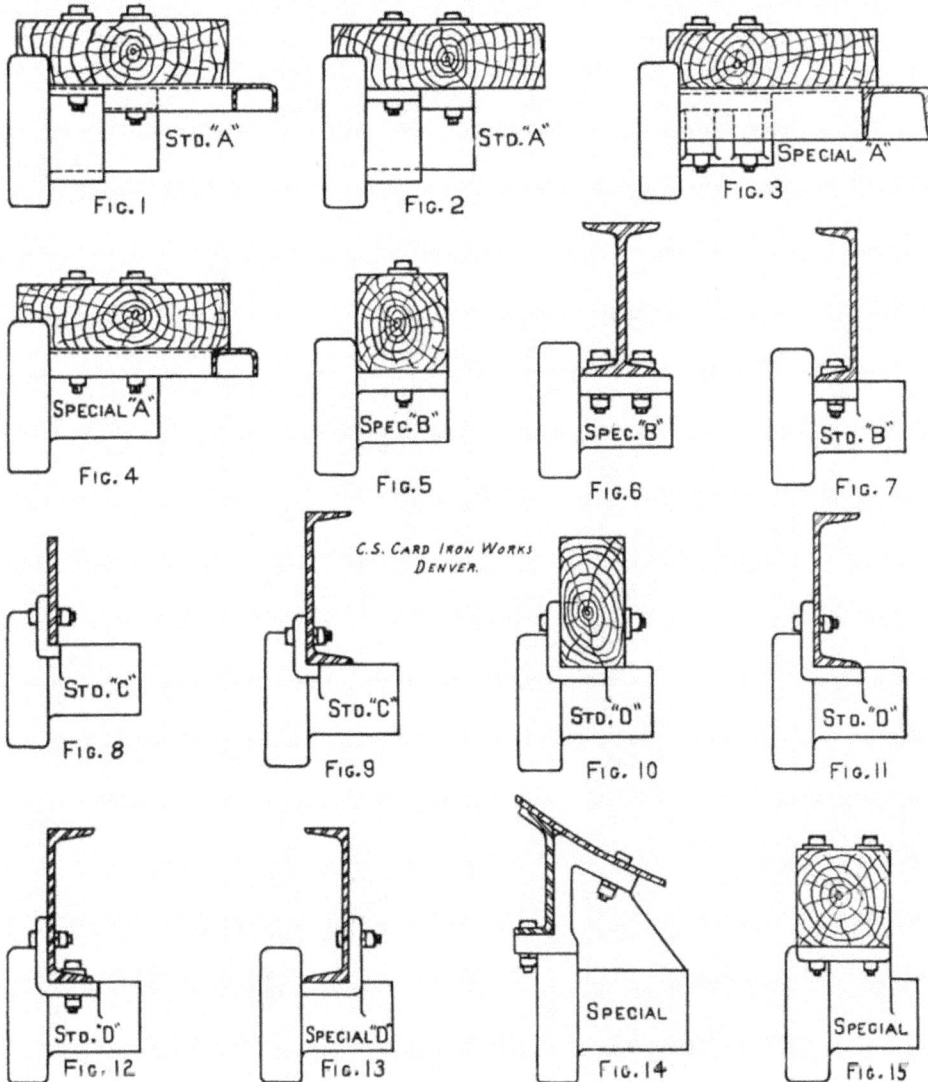

Fig. 1 — Std. "A"

Fig. 2 — Std. "A"

Fig. 3 — Special "A"

Fig. 4 — Special "A"

Fig. 5 — Spec. "B"

Fig. 6 — Spec. "B"

Fig. 7 — Std. "B"

Fig. 8 — Std. "C"

Fig. 9 — Std. "C"

Fig. 10 — Std. "D"

Fig. 11 — Std. "D"

Fig. 12 — Std. "D"

Fig. 13 — Special "D"

Fig. 14 — Special

Fig. 15 — Special

C.S. Card Iron Works Denver.

The illustrations are to show but a few of the many ways our pedestals can be used in attaching trucks to cars. By means of special design, almost any condition can be met. If you will furnish us a drawing or a dimensioned sketch of your car frame to which you wish to attach the truck, we will submit a design showing how the pedestal and truck can be applied. No obligation is entailed on your part.

See page 30 for standard types.

"CARD" AXLE AND PEDESTAL WASHERS

This washer list covers the range of sizes we regularly use in our standard equipment. We are well equipped to furnish you any size washer you require.

In ordering, be sure to specify the outside diameter, the inside diameter or hole, and the thickness.

Specifications; Prices and Average Weights in Pounds, Each

Outside Diam. Inches	Inside Diam. Inches	PRICE Thickness					WEIGHT Thickness				
		$\frac{1}{2}''$	$\frac{3}{8}''$	$\frac{1}{4}''$	$\frac{3}{16}''$	$\frac{1}{8}''$	$\frac{1}{2}''$	$\frac{3}{8}''$	$\frac{1}{4}''$	$\frac{3}{16}''$	$\frac{1}{8}''$
$6\frac{1}{2}$	$2\frac{15}{16}$	$0.46	$0.37	$0.28	$0.24	$0.20	3.75	2.80	1.87	1.40	0.93
6	$2\frac{11}{16}$	0.41	0.33	0.25	0.22	0.18	3.21	2.40	1.61	1.20	0.79
$5\frac{5}{8}$	$2\frac{11}{16}$	0.37	0.30	0.24	0.21	0.17	2.71	2.03	1.36	1.01	0.67
$5\frac{1}{4}$	$2\frac{1}{2}$	0.33	0.27	0.21	0.18	0.15	2.39	1.78	1.20	0.89	0.59
5	$2\frac{1}{2}$	0.30	0.25	0.20	0.18	0.15	2.10	1.57	1.05	0.79	0.52
	$2\frac{1}{4}$	0.30	0.25	0.20	0.18	0.15	2.23	1.67	1.11	0.83	0.55
	2	0.30	0.25	0.20	0.18	0.15	2.34	1.75	1.17	0.88	0.58
$4\frac{3}{4}$	$2\frac{1}{2}$	0.28	0.24	0.19	0.17	0.14	1.83	1.37	0.92	0.68	0.45
	$2\frac{1}{4}$	0.28	0.24	0.19	0.17	0.14	1.96	1.46	0.98	0.73	0.48
$4\frac{1}{2}$	$2\frac{3}{4}$	0.26	0.22	0.18	0.16	0.14	1.41	1.05	0.70	0.53	0.35
	$2\frac{1}{4}$	0.26	0.22	0.18	0.16	0.14	1.69	1.26	0.85	0.63	0.42
	2	0.26	0.22	0.18	0.16	0.14	1.80	1.35	0.90	0.67	0.45
$4\frac{1}{4}$	$2\frac{1}{2}$	0.23	0.20	0.16	0.14	0.12	1.32	0.99	0.66	0.49	0.33
	$2\frac{1}{4}$	0.23	0.20	0.16	0.14	0.12	1.45	1.08	0.73	0.54	0.36
	2	0.23	0.20	0.16	0.14	0.12	1.56	1.17	0.78	0.58	0.39
	$1\frac{3}{4}$	0.23	0.20	0.16	0.14	0.12	1.68	1.25	0.84	0.63	0.41
4	$2\frac{1}{4}$	0.22	0.18	0.15	0.14	0.12	1.22	0.91	0.61	0.46	0.30
	2	0.22	0.18	0.15	0.14	0.12	1.34	1.00	0.67	0.50	0.33
$3\frac{3}{4}$	$2\frac{1}{4}$	0.19	0.16	0.13	0.12	0.10	1.01	0.75	0.51	0.38	0.25
	2	0.19	0.16	0.13	0.12	0.10	1.12	0.84	0.56	0.42	0.28
	$1\frac{3}{4}$	0.19	0.16	0.13	0.12	0.10	1.24	0.92	0.62	0.46	0.31
$3\frac{1}{2}$	$2\frac{1}{2}$	0.18	0.15	0.12	0.11	0.10	0.67	0.50	0.33	0.26	0.17
	2	0.18	0.15	0.12	0.11	0.10	0.92	0.69	0.46	0.35	0.23
	$1\frac{3}{4}$	0.18	0.15	0.12	0.11	0.10	1.02	0.76	0.51	0.38	0.25
	$1\frac{1}{2}$	0.18	0.15	0.12	0.11	0.10	1.11	0.83	0.55	0.41	0.27
$3\frac{1}{4}$	$2\frac{1}{4}$	0.16	0.14	0.12	0.11	0.10	0.61	0.46	0.31	0.23	0.15
	2	0.16	0.14	0.12	0.11	0.10	0.73	0.54	0.36	0.27	0.18
	$1\frac{1}{2}$	0.16	0.14	0.12	0.11	0.10	0.92	0.69	0.46	0.35	0.23
3	2	0.14	0.12	0.10	0.09	0.08	0.55	0.41	0.28	0.21	0.14
	$1\frac{3}{4}$	0.14	0.12	0.10	0.09	0.08	0.67	0.50	0.33	0.25	0.17
	$1\frac{1}{2}$	0.14	0.12	0.10	0.09	0.08	0.75	0.56	0.38	0.28	0.19
$2\frac{3}{4}$	2	0.13	0.11	0.09	0.09	0.08	0.40	0.29	0.19	0.14	0.10
	$1\frac{1}{2}$	0.13	0.11	0.09	0.09	0.08	0.58	0.44	0.29	0.22	0.14
	$1\frac{1}{4}$	0.13	0.11	0.09	0.09	0.08	0.67	0.50	0.33	0.25	0.17
$2\frac{1}{2}$	$1\frac{1}{2}$	0.10	0.09	0.08	0.07	0.07	0.44	0.33	0.22	0.16	0.11
	$1\frac{1}{4}$	0.10	0.09	0.08	0.07	0.07	0.53	0.39	0.26	0.20	0.13
$2\frac{1}{4}$	$1\frac{1}{4}$	0.09	0.08	0.07	0.07	0.06	0.40	0.30	0.20	0.15	0.10
2	1	0.09	0.08	0.07	0.06	0.06	0.34	0.26	0.17	0.13	0.08

On account of machine set-ups there will be a minimum charge of $5.00 on small lots of washers.

SECTION "B"

CATALOG NO. 40

COAL MINE CARS

The C.S. Card Iron Works Co.
Denver, Colorado

A car assembly floor, served by overhead crane. Trunnions and nut tightening machine used in assembly of coal mine cars are shown.

Oil furnaces and bulldozer used for forming car parts such as bumpers, wheel hoods, etc.

A battery of wheel boring mills.

"CARD" COAL MINE CARS

The cars illustrated are but a few of the many different designs we have furnished. Newer and different designs are constantly being developed. Advances in metallurgy, such as alloy steels; improvement in manufacturing processes and equipment, such as arc welding as an example; changes in mining methods, as for instance mechanical loading; are to mention but a few of the things that are continually contributing to improved and changing car design.

If you are contemplating a new car, we believe we can be of assistance to you from the standpoint both of design and of manufacture. At your disposal is such information as we have collected in our many years experience (since 1892), building coal mine cars.

To draw upon, we have hundreds of car drawings of different styles of coal mine cars constructed of wood, of steel, or a combination of the two.

To intelligently submit a satisfactory car design, it is necessary that we have as much information about your operating conditions as you can furnish. Answering as many of the following questions as possible will assist us in submitting a car that will be satisfactory for your requirements.

1—Will the car be used in a shaft, slope or drift mine?

2—Will the car be hand loaded or machine loaded, or both?

3—State any preference for, or limit to, the height of the car above the rail.

4—State any preference for, or limit to, the length of the car overall.

5—State any preference for, or limit to, the width of the car overall.

6—What will be the style, or styles, of haulage:—motor, rope or animal?

7—What will be the maximum number of cars to be hauled in a trip?

8—What will the track gauge be?

9—With what type dump will the cars be used?

10—Are brakes desired?

If you have a preference for any of the following, please state it.

11—Style of Truck? (See Section "A".)

12—Wheel diameter?

13—Wheel base?

Please advise how many cars you will require.

If any of the cars we have illustrated approximate the design you have in mind, you can assist us materially by specifying by car number. The same applies to any parts such as bumper styles, style of hitchings, etc.

Car No. 344

A popular type oak body car of conventional design with center bumper and wood door. It can be used in either slope or in shaft mines with automatic cages.

A car with oak bottom, steel sides, end and door, suitable for low veins. The development of the familiar oak body car, substituting steel plate sides to obtain greater capacity and strength. The side plates are stiffened by flanging the tops.

Car No. 1558

Car No. 1058

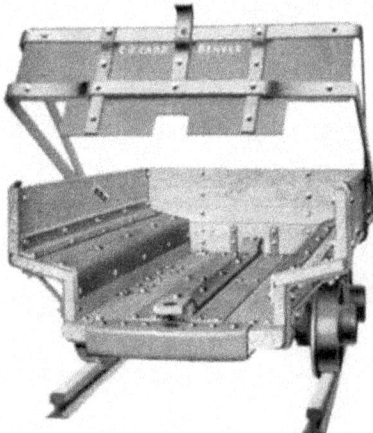

A popular type of an oak body, steel drop door car in which the cracks in the bottom and sides have been covered with steel strips to prevent the fine coal sifting through to the haulage road. The strips also stiffen the body. This strip construction is also popular on the bottoms of composite oak bottom and steel side cars.

Car No. 1128

A car with oak bottom and steel sides and ends, for use in rotary dump. The side angles stiffen the car and hold it while dumping. The brake pressure is applied through wood brake blocks on the wheels.

Car No. 1150

A substantially constructed, five binder, oak body, rotary dump car of larger capacity. Note the safety chains, as the car is used on a steep slope. The heavy center bumper covers practically the entire end.

Car No. 1256

An oak body rotary dump car with double bumpers. Note the heavy formed "Z" plate which takes the place of the lower side board. This construction materially strengthens the car, prevents the bottom sagging, and increases capacity. It is used frequently.

Car No. 1471

A rotary dump car with oak bottom and steel sides and end, designed especially for mechanical loading. Large capacity and lower center of gravity are obtained by hooding the wheels. Note the safety chains.

Car No. 1481

A larger capacity car with drop door and hooded wheels, designed for mechanical loading. Can be equipped with band brakes. The front binder is connected across the top by an angle, strengthening the door end of the car.

Car No. 1556

A rotary dump car with oak bottom and steel sides and ends. The wheels are hooded. The car is substantially braced and has proven very satisfactory in low vein coal.

Car No. 2078

An all steel car with band brakes, spring draft gear and s w i v e l coupling links. These couplings allow the cars to be rotated in the dump one at a time without uncoupling f r o m the train. The car body is completely arc-welded.

A l a r g e capacity rotary dump car with hooded wheels and steel band brake. The oak bottom is completely covered inside with steel plate. The tops of the side plates are formed to a triangular section and riveted for added strength.

Car No. 2191

Car No. 2233

A composite body car with hooded wheels, suitable for use in slope, or in shaft mines with automatic or self-dumping cages. Considerable capacity is gained by hooding the wheels.

Car No. 2283

A modern all-steel rotary dump car equipped with automatic couplings, spring draft gear, and steel band brakes on the four wheels which are hooded. The brakes are equalized and adjustable.

The car is suitable for both hand and mechanical loading. As built to 36″ gauge, and although only 36″ above the rail, it has a capacity of 125 cubic feet level full.

The interior bracing is clearly shown in the illustration. The bottom is braced equally well on the under side.

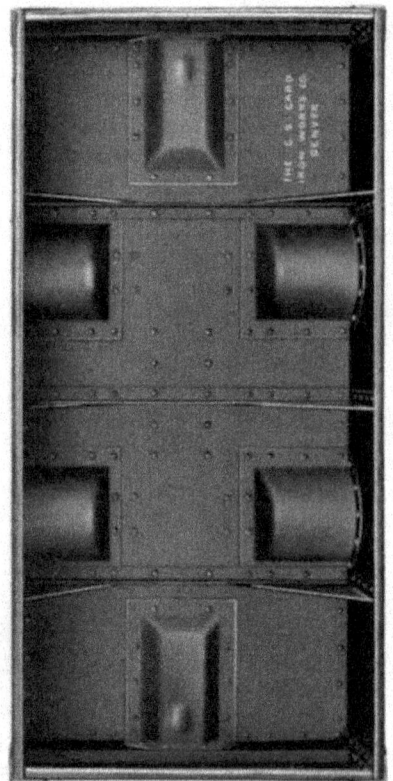

Car No. 2606

A large capacity steel body car, 206 cubic feet level full, designed for mechanical loading. It is equipped with spring draft gear and steel band brakes operating on the four wheels. Brakes are equalized on the wheels and are adjustable.

The cars are held in the rotary dump by extensions in the dump engaging in the slots in the car sides. This construction materially strengthens the car and increases capacity.

The construction is a combination of riveting and arc-welding, each used to the best advantage.

The car is well braced internally. Bottom bracing is shown in the illustration.

"CARD" MINE CAR IRONS

With our modern equipment, trained personnel, and large stock of raw materials to draw on, we can furnish you promptly with any car irons you require at reasonable prices.

In ordering or asking for quotations, please furnish us with a drawing or a dimensioned sketch of the car irons desired. Be sure to specify the weight of the material and the size and location of any necessary holes. Following are outline drawings of various styles of bumpers, drawbars and couplings which can be used as a guide.

BUMPERS

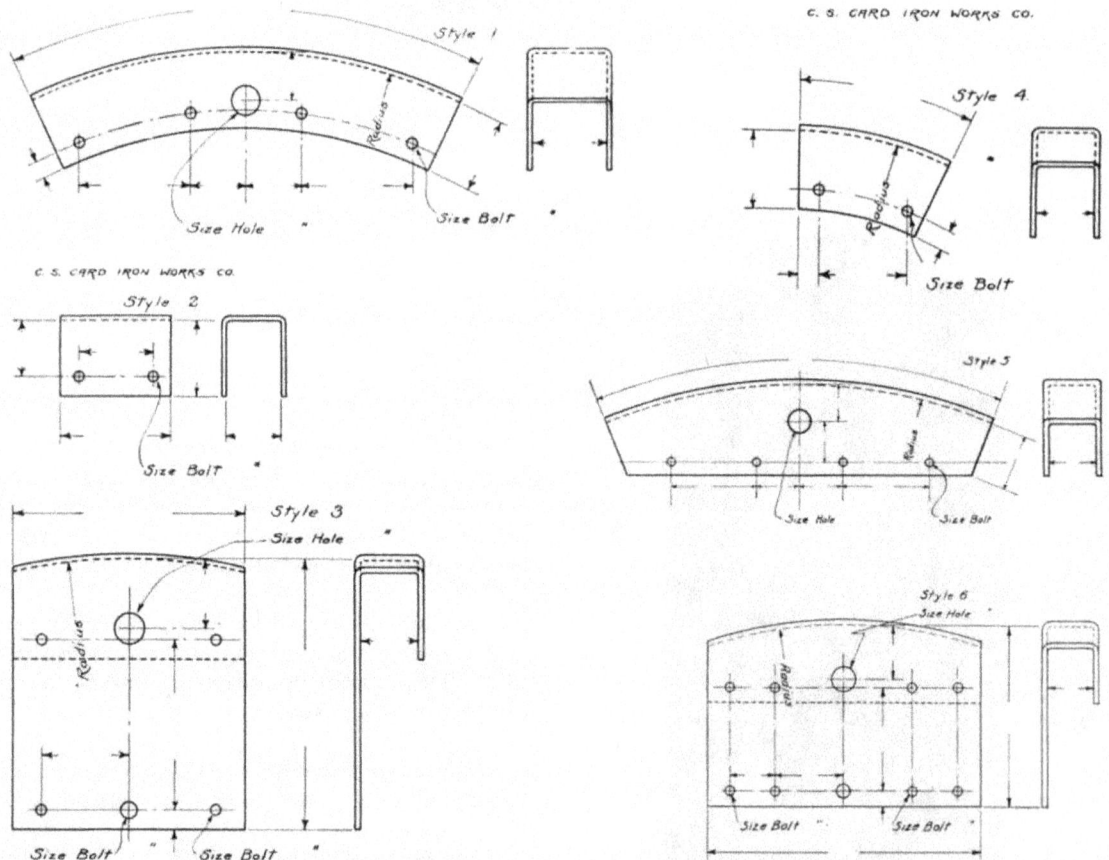

We illustrate here six of the most popular types of car bumpers. Styles No. 1, No. 3, No. 5, and No. 6 are for round or center bumper types of cars; No. 2 and No. 4 are for double bumper cars.

In ordering or asking for quotations, please advise the thickness of the steel and fill in as many of the dimensions as possible.

We can furnish any style bumper made to your specifications upon receipt of a sketch or drawing.

Additional copies of this page on application.

DRAWBARS

This page illustrates six styles of drawbars most commonly used. Styles No. 1 and No. 4 are used with double bumper cars; styles No. 5, No. 6, No. 7 and No. 8 with center or round bumper cars. We can furnish you any style of drawbar you require.

In ordering or asking for quotations, please furnish us with a rough sketch of the drawbar, being sure to furnish us the weight of the material, the size of the holes and their distances center to center.

All pin holes in the main drawbar are swaged, not drilled or punched, to give greatest possible strength.

Drawbar offsets are made while the metal is hot. This eliminates setting up unusual stresses at the bends.

Additional copies of this page on application.

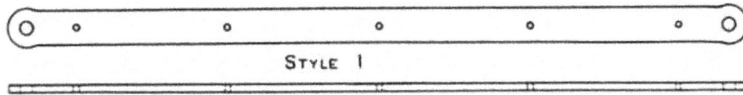

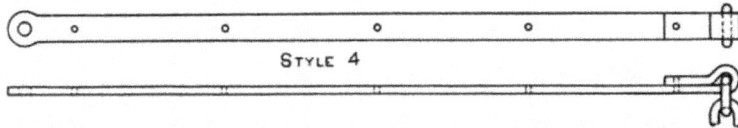

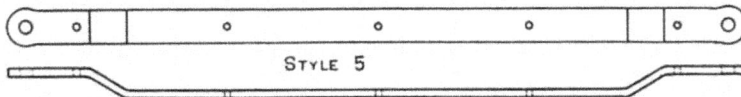

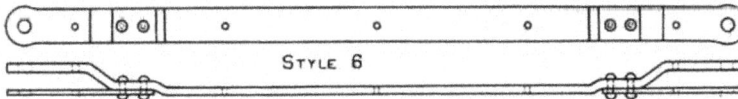

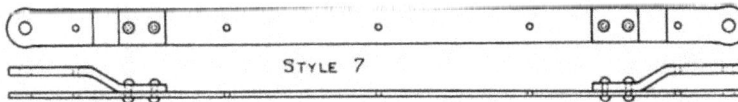

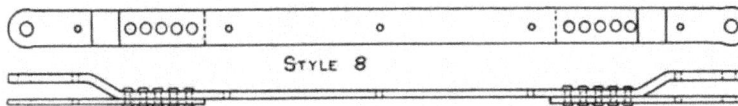

STYLE 1

STYLE 4

STYLE 5

STYLE 6

STYLE 7

STYLE 8

"CARD" MINE CAR COUPLINGS

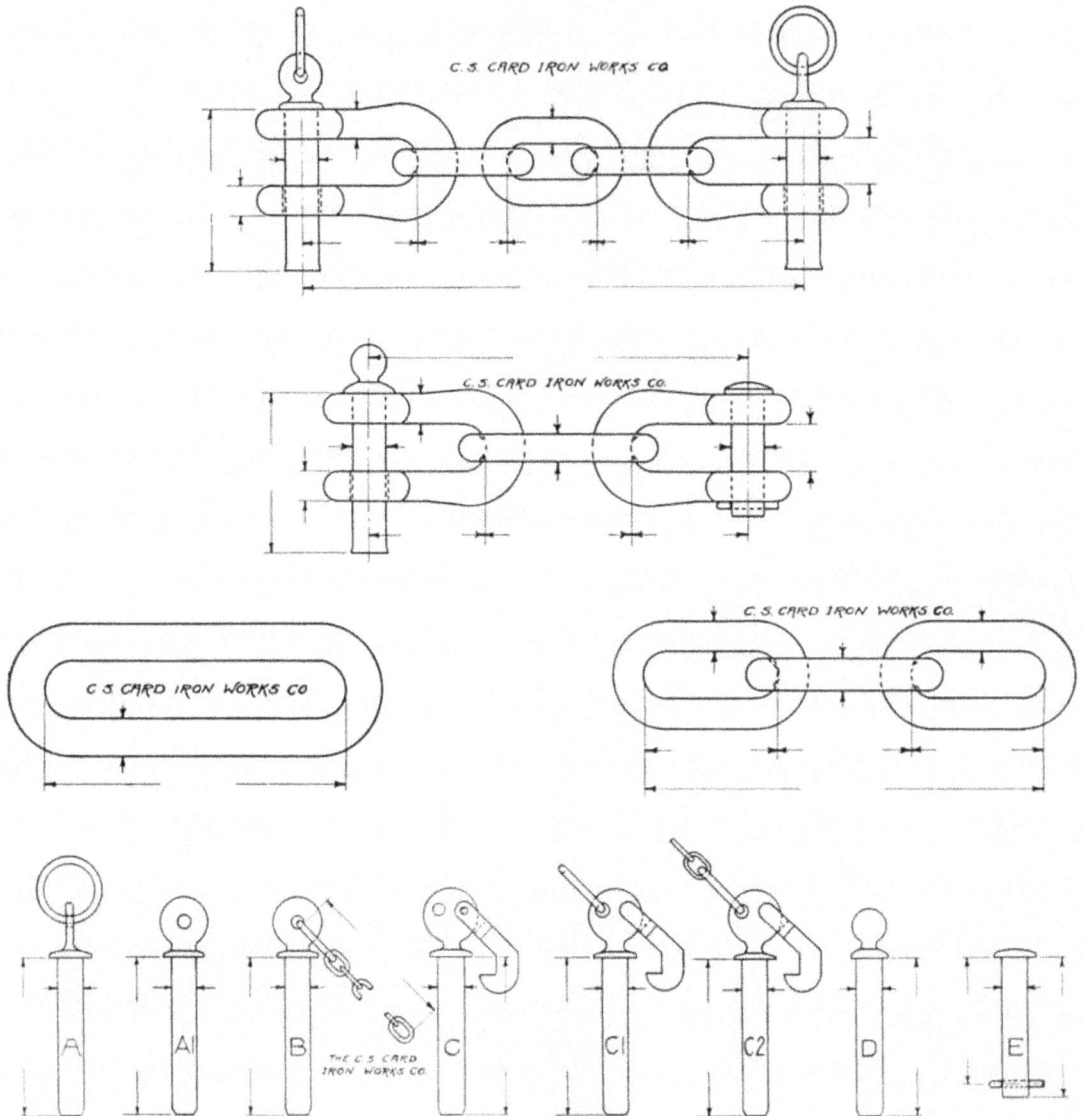

In all our couplings, we use special steels which are more uniform and superior to "Norway" iron. All clevises and pins are forged from the solid.

On account of the increased strength, we will furnish all clevises with reinforced neck or bell as illustrated, unless otherwise specified.

In ordering clevis couplings, specify the style pins desired in the clevis.

Note the style "C" pins have a safety hook or lock. The pin will not come out of the drawbar unless the hook is released by hand. Style "C" pins are very popular.

Additional copies of this specifications sheet on application.

A♠

SECTION "C"

CATALOG NO. 40

ORE AND
INDUSTRIAL CARS

The C.S. Card Iron Works Co.
Denver, Colorado

Powerful machines used for multiple punching, bending and forming of steel plates, bars, etc.

Rocker dump cars crated for export. Typical export crating is shown.

"CARD" ORE AND INDUSTRIAL CARS

The cars illustrated are but a few of the many different designs we have built. Newer and different car designs are constantly being developed. Advances in metallurgy, such as alloy steels; and improvement in manufacturing processes and equipment, such as arc-welding as an example; contribute to different and better car design.

If you are contemplating a new car, we believe we can be of assistance to you. We have hundreds of car drawings in our files covering a wide range in design.

The operating conditions govern the design of the car that can be used to the best advantage. A car that proves satisfactory and economical under one condition may not prove so under another.

So that we can intelligently submit full car information, it is necessary that we have as much knowledge of your operating conditions as possible for you to furnish. Answering as many of the following questions as you can will assist us in submitting a car that will be satisfactory for your requirements.

1—What car capacity in cubic feet level full is desired?

2—What is the weight of the material per cubic foot, as loaded?

3—What kind of material will be handled? Is it dry, wet or sticky? On what pitch will it flow when dumped?

4—What type of haulage, whether motor, rope, animal or hand?

5—What is the maximum number of cars you will handle in a train?

6—What will be the maximum speed of the train and the maximum length of the haul?

7—What is the maximum grade on which the cars will be used?

8—As the track curves will affect the wheel base, what will be the radius of the sharpest curve?

9—Will the cars be hand loaded, machine loaded, or loaded from chutes?

10—Are the cars to be dumped by hand, automatically, or by power?

11—State any preference for, or limit to, the height of the car above the rail.

12—State any preference for, or limit to, the length of the car overall.

13—State any preference for, or limit to, the width of the car overall.

14—What will the track gauge be?

15—Are brakes required?

If you have a preference for any of the following, please state it.

16—Style of truck? (See Section "A".)

17—Wheel diameter?

18—Couplers, plain link and pin type, or automatic type?

Please advise how many cars you will require. With few exceptions, ore and industrial cars are designed and built to order. The quantity fabricated at one time affects the price.

If any of the cars we have illustrated approximate the design you have in mind, please specify.

The cars illustrated are to show general designs. No attempt has been made to show all the variations and extras which can be placed on a car. Different types of drawbars, bumpers, brakes, hitchings, couplers, door mechanisms, wood lagged bottoms and liner plates, etc., can be furnished to meet individual cases.

With our modern equipped plant and trained personnel, we can furnish you at reasonable cost any style or type of car you may require.

"CARD" STANDARD ORE CARS

TYPE "Z"

Our standard ore car will meet practically all ordinary conditions where a car of this style can be used. Designed and honestly built to keep going under mine operating conditions at a minimum of attention and repairs.

The body is well balanced on the truck, is easily dumped, and the car will not leave the track when it is dumped.

The standard body is completely arc-welded. This construction makes a stronger body than a riveted one. It also reduces the weight without sacrificing strength.

Two angles across the bottom strengthen it against loading punishment, the rear angle reinforcing the bottom against dropping on the turntable.

Bodies of twenty or more cubic foot capacity are regularly furnished with one vertical reinforcing strap or band in the center of the body. This strap is not standard on cars of less than twenty cubic foot capacity.

The door overlaps the sides and has reinforcing straps at the bottom, top and sides, protecting it against being pushed inside the body as when two cars come together.

Steel bumpers, located at the strongest corners of the body, protect the door operating mechanism against damage and are used as handles when dumping.

The turntable is semi-steel, with machine finished grooves and turning surfaces, both upper and lower halves, and is grease lubricated. This construction takes most of the load off the king pin, facilitates turning and increases stability. The lower half is also machine finished to get perfect contact where it fits into the truck frame.

The finished hinge rod has ample bearing in both the turntable and the hinge bearing on the car bottom.

The steel truck frame is formed from one piece without riveted or welded corners to cause trouble. It is held rigid and in alignment by the lower half of the turntable at the top and the truck at the bottom.

These refinements result in an easier running, easier operating car with repairs held to a minimum.

Type "Z" cars are regularly equipped with "CARD" Standard Roller Bearing Truck shown on pages 3 and 4, Section "A" or "CARD" Timken Roller Bearing Truck shown on page 6, Section "A".

When used in long, heavy trains, Type "Z" cars can be equipped with "CARD" Patented Spring Drawbar Truck shown on page 8, Section "A". For lighter haulage in trains they can be equipped with Standard Solid Drawbar and chain shown on page 49.

Coupling chains on the side of the body can be furnished. We recommend against taking the "pull" through the body.

Where conditions, such as transportation problems, make it necessary or desirable, these style "Z" cars can be furnished special, knocked down for riveted construction in the field. See "Z-16" car with brake, illustrating riveted construction.

"Z" Car Standard specifications are listed on page 49.

Type "Z-16" car with standard brake. Weight applied on end of brake beam operates brake to all wheels; relieving weight releases brake.

Price, $13.00 per car extra.

Weight, 60 lbs. per car extra.

"CARD" STANDARD ORE CARS

TYPE "Z"

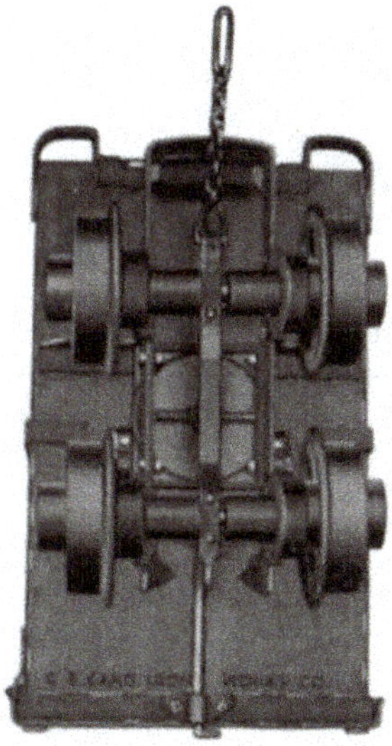

Standard "Z" car with standard plain solid drawbar with chain. The drawbar pull is continuous through the train. The drawbar is easily attached by "U" bolts to the car axles. Chain is hung on a hook on the car frame when not in use.

Price $6.40 per car extra.

Weight, 25 lbs. per car extra.

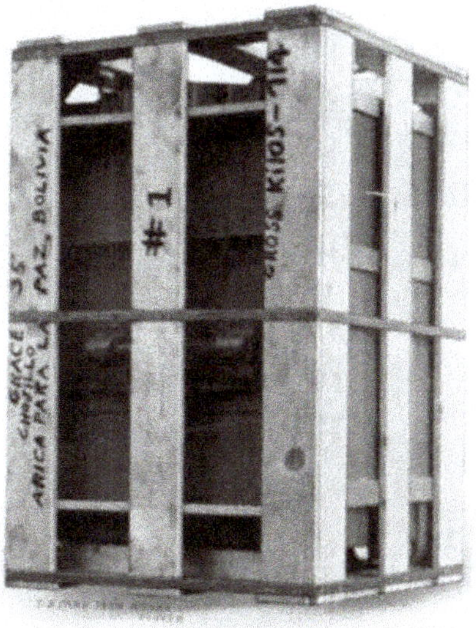

The illustration shows packing used on export shipments of Type "Z" cars. Truck and frame are packed inside the body.

"CARD" STANDARD TYPE "Z" ORE CARS WITHOUT BRAKE OR DRAWBAR

Specifications; Prices and Average Weights, in Pounds, Each

	Car No.	Capacity, Level Full, Cubic Feet	Overall Dimensions			Inside Body Dimensions			Thickness of Steel		Truck			With 'CARD' Standard Roller Bearing Truck		With 'CARD-TIMKEN' Roller Bearing Truck	
			Length	Width	Height Above Rail	Length	Width	Depth	Bottom	Sides and Door	Wheel Diam.	Axle Size	Wheel Base	Price	Weight	Price	Weight
18" Track Gauge	‡Z-12	12.0	46½"	27½"	36⅜"	42"	24"	20½"	3/16"	10 Ga.	10"	*1½"	16"	$ 78.00	660	$ 88.00	660
	‡Z-14	14.3	46½"	27½"	40⅜"	42"	24"	24½"	3/16"	10 Ga.	10"	*1½"	16"	80.00	675	90.00	675
	‡Z-16	16.2	48½"	29½"	40½"	44"	26"	24½"	3/16"	10 Ga.	10"	1¾"	16"	85.00	725	92.00	700
	‡Z-18	18.7	48½"	33½"	40½"	44"	30"	24½"	3/16"	10 Ga.	10"	1¾"	16"	88.00	750	95.00	730
	‡Z-20	20.4	52¾"	34¼"	42⅝"	48"	30"	24½"	¼"	10 Ga.	12"	2"	18"	102.00	985	111.00	925
	Z-22	23.0	58⅞"	34⅜"	42⅝"	54"	30"	24½"	¼"	3/16"	12"	2"	18"	111.00	1070	120.00	1015
	Z-24	24.9	58⅞"	34⅜"	44¾"	54"	30"	26½"	¼"	3/16"	12"	2¼"	18"	121.00	1150	130.00	1100
24" Track Gauge	Z-12	12.0	46½"	33⅜"	37½"	42"	24"	20½"	3/16"	10 Ga.	10"	*1½"	16"	$ 84.50	685	$ 94.50	685
	Z-14	14.3	46½"	33⅜"	41⅛"	42"	24"	24½"	3/16"	10 Ga.	10"	*1½"	16"	86.50	700	96.50	700
	‡Z-16	16.2	48½"	33⅜"	41¼"	44"	26"	24½"	3/16"	10 Ga.	10"	1¾"	16"	92.00	755	99.00	730
	Z-18	18.7	48½"	33½"	41¼"	44"	30"	24½"	3/16"	10 Ga.	10"	1¾"	16"	95.00	785	102.00	760
	‡Z-20	20.4	52¾"	34¼"	43⅜"	48"	30"	24½"	¼"	10 Ga	12"	2"	18"	110.00	1005	119.00	955
	Z-22	23.0	58⅞"	34⅜"	43⅜"	54"	30"	24½"	¼"	3/16"	12"	2"	18"	121.00	1110	130.00	1060
	Z-24	24.9	58⅞"	34⅜"	45½"	54"	30"	26½"	¼"	3/16"	12"	2¼"	18"	132.00	1200	141.00	1150

‡—Carried in stock for immediate shipment.
*—Cars No. Z-12 and No. Z-14 with "CARD-TIMKEN" Trucks have 1¾" axles.
Standard Z-Cars can be promptly furnished to any track gauge.

"CARD" SPECIAL TYPE "Z" ORE CARS

Many operators have found it to their advantage to have cars of our type "Z" built to meet their special operating conditions when our standard type "Z" ore cars do not. In this way they get a car which incorporates time and mine-tested design at a reasonable cost.

Any special features that are either necessary or desirable can be incorporated. Frequently the only changes required from the standard "Z" cars are in the overall measurements and size of the body.

The special type "Z" car illustrated was built for one of the large copper mining companies to meet their special conditions.

Prints of the outline drawing of type "Z" car below can be had on request. This print can be used for furnishing pertinent information that will permit us to make you a quotation or to build a car to your requirements. You can sketch any special features desired on the print we furnish.

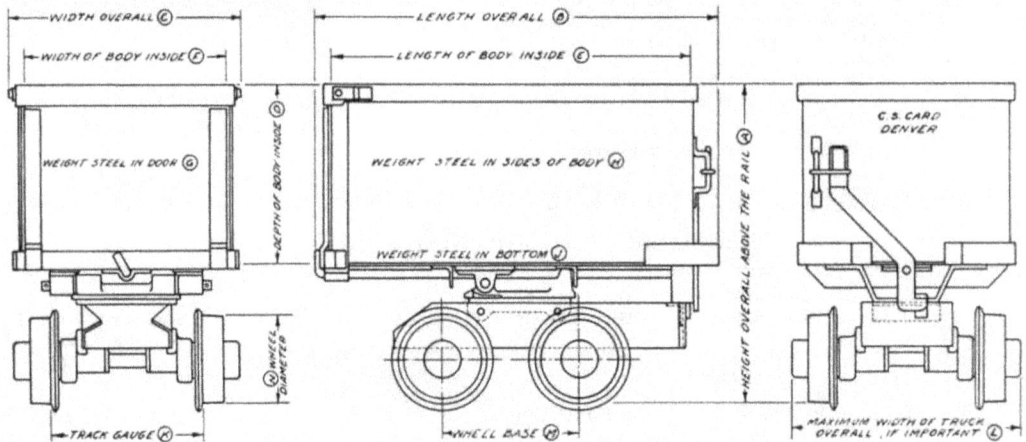

As local conditions sometimes require that cars be kept to certain specified measurements, the furnishing of data asked for below will be of assistance. Additional copies of this sheet can be had on request. With the following information we can furnish a car closely duplicating your present equipment.

A—Height over all above the rail

B—Length over all

C—Width over all

D—Depth of body inside

E—Length of body inside

F—Width of body inside

G—Weight steel in door

H—Weight steel in sides of body

J—Weight steel in bottom

K—Track Gauge

L—Maximum width of truck over all, if important

M—Wheel Base

N—Wheel Diameter

Any specifications that are optional can be left blank.

"CARD" ORE CARS

TYPE "S"

Type "S" cars are built especially for motor haulage, and are recommended for the most severe service.

They are regularly equipped with "CARD" Patented Roller Bearing Spring Drawbar Truck shown on page 8, Section "A".

Note the rounded steel bumpers on both the front and rear of the car.

A combination of door lever and foot treadle lock allows the car to be rotated without lifting the body.

The body is strongly reinforced and mounted on a grease lubricated, dirt-proof, semi-steel turntable with machined groove that takes part of the load from the king pin when the body is rotated.

Built to order only. Specifications and prices on request.

"CARD" SCOOP CARS

This standard scoop car with its solid body can be used to advantage in handling fine or wet materials such as concrete.

The arc-welded body is mounted on a grease lubricated, dirt-proof, semi-steel turntable with machined groove that takes part of the load from the king pin when body is rotated. Body is locked by a treadle catch at rear.

Regularly equipped with "CARD" Roller Bearing Truck shown on page 4, or "CARD" Timken Roller Bearing Truck shown on page 6, Section "A", and furnished for 18" gauge, unless otherwise specified. Carried in stock for immediate shipment.

Specifications; Prices and Average Weights in Pounds, Each 18" Track Gauge Car

Car No.	Capacity, Cubic Feet	Overall Dimensions			Inside Body Dimensions			Thickness of Steel		Truck			With 'CARD' Standard Roller Bearing Truck		With 'CARD-TIMKEN' Roller Bearing Truck	
		Length	Width	Height Above Rail	Length	Width	Depth	Bottom	Sides	Wheel Diam.	Axle Size	Wheel Base	Price	Weight	Price	Weight
K-12	11.6	51¼"	27⅜"	37¹³⁄₁₆"	48"	24"	21"	³⁄₁₆"	10 Ga.	10"	*1½"	16"	$85.00	600	$95.00	600
K-18	18.0	55¼"	31"	43"	52"	30"	24½"	¼"	10 Ga.	12"	1¾"	16"	$99.00	780	106.00	740

*—K-12 car with "CARD-TIMKEN" Truck has 1¾ axles.

"CARD" ROCKER DUMP CARS

CAR NO. 2201

Car No. 2201 was built for a large Mexican mining company and is quite typical of the design of rocker dump cars used on the narrower gauges and for limited overall widths. Body is carried on heavy cast steel rockers and rocker stands, which are covered by a plate to protect against spilled material lodging therein.

Rocker dump cars are adaptable to a large number of conditions and are popular in mines, quarries, and the contracting field. While their principle of operation is the same in all instances, the design and the capacity is largely governed by the limiting overall dimensions and the track gauge.

Rocker dump cars, when properly designed, are easily dumped to either side and easily returned to upright position.

Various methods are used to lock the body in upright position. These locks usually operate automatically to stop the body in the upright position when it is returned from the dumped position.

The bodies are tight (no doors) and there are few operating parts to get out of order and cause trouble and repair expense. There are no openings through which material can leak along haulage ways.

The cars shown on this page are very popular in twenty to forty cubic foot capacities on 18" to 24" gauges.

Rocker dump cars are built on order to meet operating requirements. Please refer to page 47 of this section.

CAR NO. 2467

Car No. 2467 shows a 40 cubic foot capacity car for 24" gauge. The locking mechanism, located on the center of the car body bottom and on the center of the car frame is plainly shown. The body is arc-welded construction.

"CARD" ROCKER DUMP CARS

CAR NO. 2524

Car No. 2524 is typical of rocker dump car design when built for the wider track gauges and to low height above the rail. Car shown is very satisfactory for hand loading. Large capacity is obtained by loading well above the car sides. The body is completely arc-welded.

CAR NO. 1653

Rocker dump car No. 1653 is a 36 cubic foot capacity car for 24" gauge. The body is riveted construction.

CAR NO. 1203

Car No. 1203 was built for use in a clay pit. Note the brake, which operates on all four wheels. Design is typical of rocker dump cars where the height and width is not too limited.

The automatic body locking mechanism is located at the four corners.

"CARD" GABLE BOTTOM CARS

CAR NO. 2219

Car No. 2219 is a large capacity car with double trucks. It is equipped with gear mechanism for operating the doors.

Gable bottom cars are built with a stationary inverted "V" bottom, set at an angle on which the material will flow to both sides when the doors are opened.

Gable bottom cars can be built with large capacity on the narrower track gauges where operating conditions permit a long body and the use of double trucks as Car No. 2219 illustrated.

There are numerous types of door operating mechanisms, several of which are shown in the illustrations. The doors are subject to severe usage and should be strongly braced to prevent warping and to hold them in alignment.

Gable bottom cars are built on order to meet operating requirements. Please refer to page 47.

CAR NO. 2091

Gable bottom car No. 2091, while showing little reinforcing, has an exceptionally strong door. Internal bracing is used. The door hinge extends the full length. This car is an instance where, by redesigning, we increased the carrying capacity fifteen percent without increasing the overall dimensions.

"CARD" GABLE BOTTOM CARS

CAR NO. 2072

Car No. 2072 was built for 18″ gauge with a capacity of 40 cubic feet and comparatively low in height above the rail. Equipped with automatic couplers and spring draft gear.

CAR NO. 1343

Car No. 1343 is a small gable bottom car, 18 cubic feet capacity, 18″ gauge. Note the solid bumper. Car is equipped with "CARD" Patented Roller Bearing Spring Drawbar Truck shown on page 8, Section "A".

CAR NO. 1666

Gable bottom car No. 1666 was built for heavy duty. Note the heavy rail bracing on the door. Car has a capacity of 104 cubic feet and operates on 30″ track gauge.

"CARD" ONE-WAY SIDE DUMP CARS
GRANBY TYPE

CAR NO. 1640

An example of a large car built for the most severe service. Car No. 1640 has a capacity of 160 cubic feet and operates on 36″ gauge. Bottom, sides and door are all oak lagged and covered with heavy steel plate. Car is equipped with air brakes. Note the heavy rail bracing.

Granby type one-way side dump cars are operated by running the train past a dump or tripping block, each car being dumped independently and without uncoupling. A dump block is shown with the illustration of Car No. 2561, while Car No. 2397 is shown in a dumped position when the dump roller is at the top of the block.

We can recommend the "Granby" car when the operating conditions permit its use. A large number of cars can be dumped in a short period. No labor at the unloading point is required to dump the cars.

Since the locomotive usually passes the dump block, there is a relationship between the locomotive, car and block. Consequently, it is advisable to furnish us with a dimensioned end or sectional view of the locomotive to be used so that we can design car for clearances and operation.

Granby type cars are built on order to meet operating requirements. Please refer to page 47.

CAR NO. 2561

Note the smooth, streamlined appearance of Car No. 2561. Body corners are rounded to keep the muck from packing in them.

Note the truck construction with outside spring mounting. This truck is shown in detail on page 17 of Section "A".

The entire car was fabricated from special alloy high strength steels.

Car has a capacity of 52 cubic feet, not including the side extension. Operates on 18″ track gauge.

Patents Pending

"CARD" ONE-WAY SIDE DUMP CARS

GRANBY TYPE

A special design "Granby" type car with a retractable dump roller feature that gives increased capacity where width is limited. This feature also permits selective dumping. The dump roller can easily and quickly be pulled into position for dumping and returned to the recess in the body.

The d r a w h e a d can be quickly swung to the one side for close coupling of mucking machine, allowing maximum loading of the car.

The car body and frame is fabricated entirely from low alloy high tensile steels. Note the rounded body corners to keep muck from packing therein. Trucks are spring mounted.

CAR NO. 2771

Patents Pending

CAR NO. 2364

Car No. 2364 was built for 30" gauge and has a capacity of 50 cubic feet exclusive of the door and side extensions above the ends. These extensions naturally increase the actual carrying capacity of the car and also prevent much spillage in loading from side chutes. Car has a renewable bottom liner plate of abrasion resisting alloy steel.

The illustration of the 53 cubic feet capacity Car No. 2397 in dumped position shows clearly the roller bearing dump roller used to raise the body and the "hold down" device used to engage in the dump block and to prevent the car truck and frame from leaving the track, should there be any tendency for them to do so.

CAR NO. 2397

"CARD" ONE-WAY SIDE DUMP CARS

CAR NO. 1767

Car No. 1767 is a 34 cubic foot capacity car, built on a 24" track gauge. The body is dumped and the door opened and closed by the lever arrangement at the end.

CAR NO. 1637

Car No. 1637, while of the same general design as Car No. 1767, has a different dumping and door operating mechanism and locking device. Internal body bracing is used.

CAR NO. 1294

A rocker or roller principle is employed in dumping the body of Car No. 1294 instead of dumping about a pivot as in the two cars above. Through levers, the door is automatically opened as the car body is dumped, and automatically locked as the car body is returned.

One-way side dump cars are built on order to meet operating requirements. Please refer to page 47.

"CARD" ROTARY DUMP CARS

CAR NO. 2078

Car No. 2078 was designed for handling potash salts. It operates on a 42" track gauge and has a capacity of 101 cubic feet. It is equipped with band brakes. The body is completely arc-welded.

The solid body and absence of doors and other moving parts on rotary dump cars permit the most substantial design. No other type of car can be so strongly constructed, factors such as the weights of materials and the capacity of the cars being the same. Relationship of tare to pay load is extremely low. Because the car body has no moving parts, repairs are naturally at a minimum. The tight body prevents spilling of materials along the haulage ways.

With limited dimensions of overall width and height above rail, maximum capacity and low center of gravity can be obtained as the body can be set very close to the truck and wheels. Sometimes the wheels are hooded. Shape of the body in section is subject to variations. Two such shapes are shown.

Cars can be dumped either as a single uncoupled unit, or by the use of swivel couplings can be dumped without uncoupling in trains. Car No. 2078 uses a swivel link connection and Car No. 1651 uses swivel type automatic coupler; both are spring cushioned.

Please refer to pages 88 to 90, Section "E", showing a number of rotary dumps which we manufacture.

Rotary dump cars are built on order to meet operating requirements. Please refer to page 47.

CAR NO. 1651

Built to a "U" shape, Car No. 1651 operates on 30" track gauge and has a capacity of 60 cubic feet. The body is built with the bottom protected with an oak cushion covered with a renewable liner plate.

"CARD" FLAT TOP OR PLATFORM CARS

We illustrate two extremes of platform cars. We build these cars in a number of different sizes and designs and with both steel and wood tops.

Platform cars are built on order to meet operating requirements.

Please refer to page 47.

"CARD" INCLINE CARS

CAR NO. 1361

We build many types of cars for handling materials on inclines, in addition to skip cars as shown on page 62.

Note that on Car No. 1361 the door is unlocked automatically by tripping a lever with roller at the dumping point, with car bottom on such angle as will allow the material to flow out.

Incline cars are built on order to meet operating requirements. Please refer to page 47. Also advise the grade of the incline on which cars are to operate.

"CARD" HOPPER BOTTOM CARS

Hopper bottom cars are built in many sizes and designs, with a number of different types of bottom gates, usually for discharging material between the rails. Gates of the slide type operated by levers, and gates operated by rack and pinion with hand wheel are more frequently used. Other types are available.

Car No. 933 is equipped with a lever operated slide gate and with handles for hand tramming.

Hopper bottom cars are built on order to meet operating conditions. Please refer to page 47.

CAR NO. 933

"CARD" BOILER ROOM WAGONS

Boiler room wagons are used principally for handling coal or ashes. When the door is dropped, it is slightly above the bottom so that a shovel will not catch on the bottom.

These wagons can be equipped with regular car trucks with flanged wheels, either plain or roller bearings, for operating on steel rail track.

List covers wagons with flat tread plain bearing wheels. They can be specially equipped with roller bearing wheels. They can be built to special sizes. Prices on application.

Specifications; Prices and Average Weights in Pounds, Each

Capacity, Cubic Feet	Overall Dimensions			Inside Body Dimensions			Thickness of Steel		Truck		Price	Weight
	Length	Width	Height above Floor	Length	Width	Depth	Bottom	Sides and Door	Wheel Diam.	Axle Size		
24	59½"	37¼"	39½"	48"	36"	24"	$\frac{3}{16}$"	10 Ga.	14"	1½"	$130.00	600
30	71½"	37¼"	39½"	60"	36"	24"	$\frac{3}{16}$"	10 Ga.	14"	1½"	139.00	650
36	83½"	37¼"	39½"	72"	36"	24"	$\frac{3}{16}$"	10 Ga.	14"	1½"	150.00	750

"CARD" SKIP CARS

CAR NO. 1623

We furnish skip cars to meet the operating conditions. Skip No. 1623 is a Style No. 2 skip without a top. Usually they have a top.

Please furnish the capacity desired, weight per cubic foot of the material to be handled, track gauge, incline on which the skip will operate, any limiting dimensions, together with any special operating conditions.

We recommend skip cars be equipped with roller bearing trucks. The truck we regularly use is shown on page 16, Section "A". Other styles of trucks can be furnished.

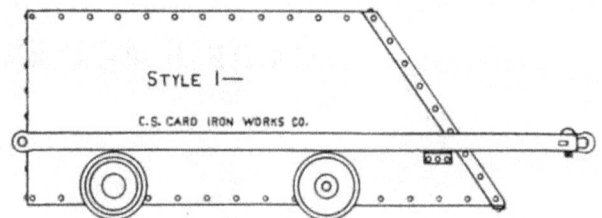

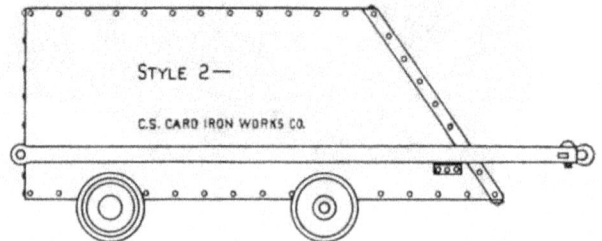

You will note the difference between Style No. 1 and No. 2 is in the mounting of axles and wheels. Where it is possible to use it, we recommend the Style No. 2 skip. The construction is more substantial. Style No. 1 permits a little greater capacity in a given slope cross section.

STYLE 1—

C.S. CARD IRON WORKS CO.

STYLE 2—

C.S. CARD IRON WORKS CO.

When required, skips will be furnished with valve or door for handling water.

See page 87, Section "E" for skips for use in vertical shafts.

CAR NO. 2102

Car No. 2102 is a skip car of different design with an open top and a door. It is lowered with the door end down the incline. At the dump point the door is automatically unlocked by tripping the levers with the rollers; and with the bottom on such a pitch that the material can flow out.

SECTION "D"
CATALOG NO. 40

ROPE HAULAGE
EQUIPMENT
ROLLERS, SHEAVES, Etc.

The C.S. Card Iron Works Co.
Denver, Colorado

"CARD" ROPE HAULAGE EQUIPMENT

We have been manufacturers of rope haulage supplies since 1892. The equipment listed in this section is the result of this long experience.

This standard equipment should fulfill almost any of your requirements. If not, take up your requirements with us, as we have a large number of special design sheaves and rollers, which are not listed as standard or shown in this section.

We carry in Denver stock practically all of the standard equipment listed in this section. Prompt shipment can be made on all items listed.

Note that the wearing surfaces of most of the sheaves and rollers are chilled, which greatly increases their life. They will not cut or wear so quickly.

All chilled sheaves and rollers are made from "CARD'S" Semi-Steel metal and are heat-treated.

Our rope haulage equipment is all liberally proportioned to stand the service required of it. Do not confuse it with the ordinary equipment of the kind made in jobbing foundries, to sell at any price.

We recommend that you use, wherever possible, the roller bearing equipment with grease lubrication. Their advantages:—the ease with which they turn—particularly in starting to turn and getting up to the rope speed—and the very infrequent and more satisfactory lubrication justifies their use. They wear enough longer to justify their additional cost. The trade has appreciated their advantages and by far the greatest part of our output is roller bearing equipped.

We suggest the use of the largest size sheave or roller that your operating conditions permit. It will prove economical, as you will obtain better wear and longer life from both the roller or sheave and the rope.

Sheaves and rollers can be furnished with plain bronze or self-lubricating graphite filled bronze bushings, the latter requiring no oil or grease lubricating.

We suggest the use of "CARD" Standard Rope Haulage Equipment wherever possible. You get the advantage of prompt shipment from stock, prompt replacement of parts, and thoroughly developed design.

"CARD" STANDARD ROPE SHEAVES

HEAVY PATTERN ONLY

We make all our standard rope sheaves from our high quality wheel metal, in heavy pattern only, designed for heavy duty.

Unless otherwise specified, we will furnish sheaves as follows for the shaft shown in the table:

4″ to 12″ sheaves with solid or web center and with one set screw.

16″ to 30″ sheaves of ribbed spoke pattern and with two set screws each.

36″ and larger sheaves of rib spoke pattern and key-seated.

Standard sheaves 16″ and larger can be made with a web center, if necessary, and where the rope pull or service requires it. Prices on application.

Prices for sheaves with special bores, special hubs, or with keyways other than specified, furnished on receipt of specifications.

All sizes of standard plain groove sheaves are carried in stock for immediate shipment.

We recommend the use of turned groove sheaves, especially when the sheaves are used in hoisting.

Turned groove sheaves are carried in stock in the rough and furnished on order. Specify the rope diameter to be used so we can turn the groove to properly support the rope.

Specifications; Prices and Average Weights in Pounds, Each

| Sheave Size | Diameter | | Maximum Rope Diameter ** | Shaft Size | Price, Plain Groove | | Add for Turned Groove | Weight, Plain Groove | |
	Groove *	Outside			Sheave Only	Sheave Shaft and Boxes ***		Sheave Only ****	Sheave Shaft and Boxes
4″	4″	6″	7/8″	3/4″	$1.55	$4.10	$0.35	7	18
5″	5″	7 1/4″	7/8″	3/4″	1.65	4.15	0.35	9	20
6″	6″	8 1/2″	1″	1″	1.80	4.45	0.55	15	27
7″	7″	9 1/2″	1″	1″	2.05	4.75	0.75	17	29
8″	8″	10 1/2″	1″	1″	2.25	5.00	0.85	19	30
9″	9″	11 5/8″	1″	1″	2.35	5.10	1.00	21	32
10″	10″	12 3/4″	1″	1 3/16″ x 8 1/2″	2.60	5.45	1.10	27	40
12″	12″	14 3/4″	1″	1 7/16″ x 10 1/2″	3.85	7.90	1.25	33	53
16″	16″	18 7/8″	1″	1 7/16″ x 11″	5.10	9.20	1.45	47	67
18″	18″	21″	1 1/8″	1 11/16″ x 11 1/2″	6.40	11.25	1.65	58	82
20″	19 1/4″	23 3/4″	1 1/8″	1 11/16″ x 12″	8.50	13.40	2.15	85	110
24″	23 1/4″	27 3/4″	1 1/8″	1 15/16″ x 13 1/2″	11.10	17.10	2.50	114	149
30″	29″	34″	1 1/8″	2 3/16″ x 15″	16.00	23.25	2.75	155	201
36″	35 1/4″	40″	1 1/8″	2 7/16″ x 16 1/2″	20.25	31.64	3.00	221	283
42″	41 1/4″	46 1/2″	1 1/4″	2 15/16″ x 24″	33.20	49.85‡	13.60	345	453‡
48″	47″	52 3/4″	1 1/4″	2 15/16″ x 24″	37.30	54.00‡	14.25	450	558‡

*—Groove diameter of turned groove sheaves is approximately 1/4″ larger than list.

**—For plain groove sheaves. Turned groove sheaves 1/8″ larger.

***—Plain cast iron boxes.

"CARD" BICYCLE SPOKE SHEAVES

We recommend the use of these sheaves on head frames, for hoisting in shafts or as bull wheels where there is an exceptionally heavy strain on the sheave. They are lighter, for the same strength, than solid cast sheaves which are always subject to some internal stresses.

We make only a heavy pattern which we know will stand up. The very liberal depth of rim insures strength and stiffness, as well as long life. Hub and rim are "CARD" semi-steel wheel metal.

The spokes are all of a special analysis steel which our experience indicates is best for this use.

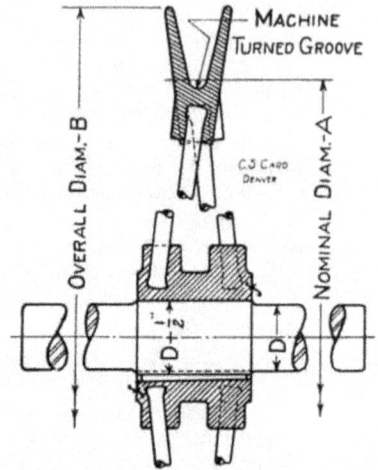

All our sheaves are made with the groove turned for the rope. Always specify the rope diameter to be used, as we machine the groove to properly support the rope. The machined groove sheave supports the rope, runs true, and requires less attention and repairs than a lined sheave.

Note the construction where the sheave is keyed to the shaft. The enlarged shaft at this point prevents it being weakened by the key-seat. The shaft is railway axle steel specifications and is finished all over. Sheaves are keyed and pressed on the shaft.

Extra heavy rigid ring oiling journal boxes are furnished and are included in the following table. Drawing showing the setting of these sheaves on request.

While we list the "Sheave Only", we recommend purchasing it "Keyed to the Shaft" at the factory where we have all facilities for making proper fit.

We can furnish these sheaves with special length of hub, special bore and any size shaft to meet your special specifications when necessary.

All listed sizes are carried in stock in the rough and finished to specifications. Prompt shipment.

Specifications; Prices and Average Weights in Pounds, Each

Sheave Size		48"	54"	60"	72"	84"	96"	108"
Nominal Diameter	"A"	48⅜"	54"	60"	72"	84"	96"	108"
Outside Diameter	"B"	52¾"	64"	70"	82"	96"	108"	120"
Shaft Diameter	"D"	3¹⁵⁄₁₆"	4⁷⁄₁₆"	4⁷⁄₁₆"	4¹⁵⁄₁₆"	4¹⁵⁄₁₆"	5⁷⁄₁₆"	5¹⁵⁄₁₆"
Shaft Length		38½"	41"	41"	47"	49"	51"	53"
Hub Length		7½"	8"	8"	10"	12"	14"	16"
Sheave Only, bored, turned and keyseated	Price	$98.25	$126.40	$141.90	$173.20	$199.50	$242.50	$281.60
	Weight	656	1067	1399	1705	2358	3135	3800
Sheave, keyed to shaft	Price	$127.80	$163.40	$178.90	$219.15	$249.50	$298.00	$347.70
	Weight	796	1252	1584	1972	2638	3490	4238
Sheave keyed to shaft and with two ring-oiling boxes	Price	$185.50	$227.50	$243.00	$290.00	$320.00	$375.00	$430.00
	Weight	1106	1637	1969	2441	3107	4019	4347

"CARD" CONCAVE CHILLED FACE KNUCKLE SHEAVES

PLAIN AND ROLLER BEARING

These sheaves are designed especially for use on knuckles.

The sheave face is chilled to minimize rope wear.

Use the largest diameter possible for longest life of both rope and sheave.

All sizes are carried in stock for immediate shipment.

Plain Bearing

The plain bearing sheaves are set-screwed to the shaft which runs in solid cast iron journal boxes.

Roller Bearing

The roller bearing sheave is equipped with solid, straight, special analysis steel roller bearings, as used in our mine car wheels, and of ample size and length to carry a heavy load. It is grease lubricated. It cannot be set up wrong, binding the bearings. Shaft is a special analysis steel.

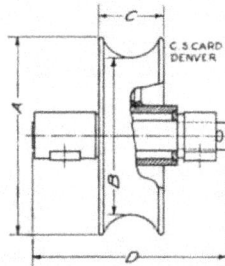

Plain and Roller Bearing

Specifications; Prices and Average Weights in Pounds, Each

Sheave Number		Plain Bearing						Roller Bearing		
		118	119	121	123	127	131	242	254	266
Outside Diameter	"A"	$10\frac{1}{2}''$	$10\frac{3}{4}''$	$15''$	$15''$	$21''$	$27''$	$15''$	$21''$	$27''$
Groove Diameter	"B"	$8\frac{1}{2}''$	$8\frac{1}{2}''$	$12''$	$12\frac{1}{8}''$	$18''$	$24''$	$12''$	$18''$	$24''$
Face Width	"C"	$4\frac{1}{2}''$	$5\frac{1}{2}''$	$3\frac{5}{8}''$	$5''$	$5\frac{1}{4}''$	$5\frac{1}{4}''$	$5''$	$5\frac{1}{4}''$	$5\frac{1}{4}''$
Length Overall	"D"	$13''$	$13\frac{3}{4}''$	$12''$	$15\frac{1}{4}''$	$16\frac{1}{4}''$	$16\frac{1}{4}''$	$14\frac{7}{8}''$	$14\frac{7}{8}''$	$14\frac{7}{8}''$
Shaft Size		$1\frac{7}{16}''$	$1\frac{7}{16}''$	$1\frac{7}{16}''$	$1\frac{15}{16}''$	$2\frac{3}{16}''$	$2\frac{3}{16}''$	$2\frac{3}{16}''$	$2\frac{3}{16}''$	$2\frac{3}{16}''$
Sheave, shaft	Price	$6.60	$6.90	$8.15	$10.75	$15.10	$18.75	$14.70	$17.90	$21.70
and boxes	Weight	51	60	63	91	136	169	102	137	170
Sheave	Price	$3.30	$3.65	$5.30	$6.25	$10.00	$14.60	$7.25	$10.90	$15.95
only	Weight	30	39	42	54	91	124	58	93	126

"CARD" STYLE "A" CURVE SHEAVES

Our Style "A" Plain Bearing Curve Sheaves, commonly called forty-five degree sheaves, have been on the market so long that most operators are familiar with them. Set inside the track, they will be found useful on curves or in pulling out of entries. The groove of the sheave wheel is chilled to minimize wear and sheave is made of "CARD" semi-steel wheel metal.

All sizes carried in stock for immediate shipment. We recommend that the special angle curve sheaves on page 69 be substituted for Style "A" sheaves wherever conditions permit. They are easier on the rope.

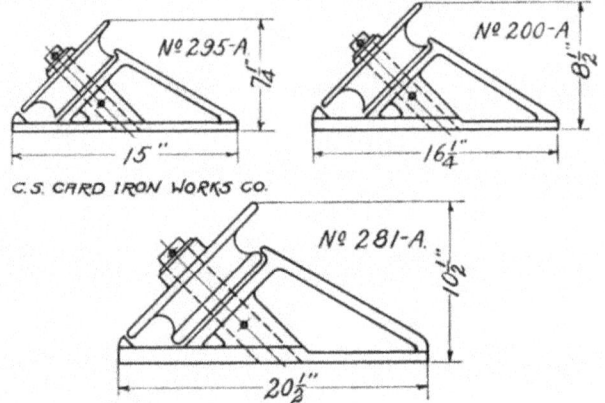

Specifications; Prices and Average Weights in Pounds, Each

Sheave Number	295–A		200–A		281–A	
Outside Diameter Groove Diameter	8" 4¾"		9½" 5½"		12" 6½"	
	Price	Weight	Price	Weight	Price	Weight
Curve Sheave Complete	$4.00	36	$4.60	52	$8.30	99
Sheave Only	1.25	12	1.65	23	2.90	43
Base and Shaft	2.95	23	3.15	28	5.90	55

"CARD" SPECIAL ANGLE CURVE SHEAVES

PLAIN AND ROLLER BEARING

These sheaves can be used to advantage many places and in design fall between our Style "A" and Style "B" curve sheaves.

The sheaves have a web center, reinforced by ribs, and the rope wear on the groove is reduced to a minimum by the chilled face.

We suggest the use of these sheaves in preference to our Style "A" sheaves wherever possible. Sheaves may be set inside the rails on the wider gauges.

The roller bearing sheaves are grease lubricated through a large reservoir direct to the bearings, and regularly furnished with a standard grease fitting. The caps have guards that prevent the rope from catching on it or the shaft. Shafts are a special analysis axle steel.

Both plain and roller bearing sheaves carried in stock for immediate shipment, in the sizes listed.

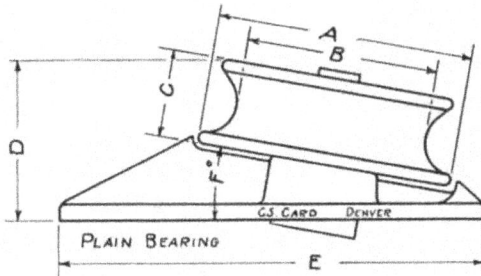

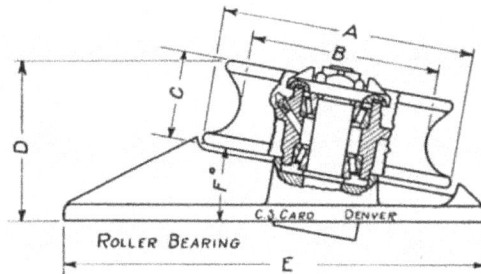

Specifications; Prices and Average Weights in Pounds, Each

	Plain Bearing				Roller Bearing			
Sheave Number	1407		1117		17-7		10-9	
Outside Diameter "A"	9″		12″		9″		12″	
Groove Diameter "B"	7¼″		9″		7¼″		9″	
Width of Face "C"	3″		4″		3″		4″	
Overall Height "D"	6⅝″		7½″		6⅝″		7½″	
Overall Length "E"	15″		20″		15″		20″	
Angle "F"	17 Degrees		10 Degrees		17 Degrees		10 Degrees	
	Price	Weight	Price	Weight	Price	Weight	Price	Weight
Curve Sheave Complete	$6.60	53	$8.40	83	$13.10	55	$15.60	88
Sheave Only, Without Bearings	2.20	20	3.00	39	4.50	21	5.60	41
Base and Shaft	4.65	32	5.80	43	5.95	30	6.65	41

"CARD" STYLE "B" CURVE SHEAVES

PLAIN BEARING

Our Style "B", or Horizontal Curve Sheave is designed for use on curves or pulling out of entries, and is usually set outside the track.

The groove of the sheave is chilled to minimize wear. Sheave is made of "CARD" semi-steel wheel metal and heat-treated.

For best results we recommend the use of the largest possible size, giving longer life to both rope and sheave.

Sheaves fitted with graphite filled self-lubricating bronze bushings made on order only. Prices on application.

See page 71 for Style "B" Roller Bearing Sheaves.

All sizes plain bearing sheaves are carried in stock for immediate shipment.

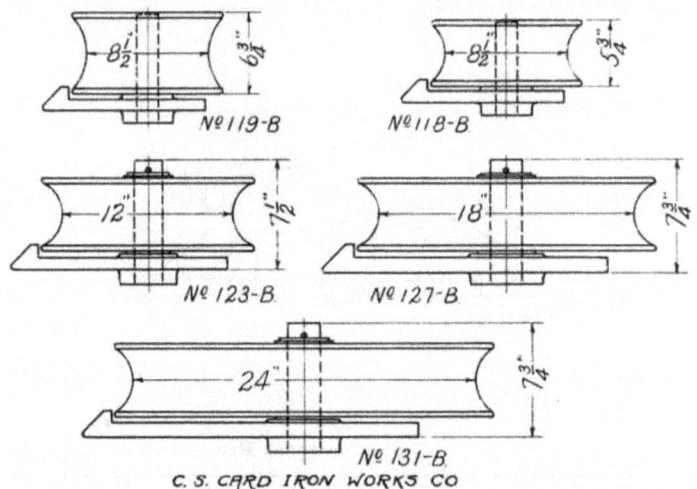

Nº 119-B

Nº 118-B

Nº 123-B

Nº 127-B

Nº 131-B

C. S. CARD IRON WORKS CO

Specifications; Prices and Average Weights in Pounds, Each

Sheave Number		118–B	119–B	123–B	127–B	131–B
Outside Diameter of Sheave Width of Face of Sheave		10½″ 4½″	10½″ 5½″	15″ 5″	21″ 5¼″	27″ 5¼″
Curve Sheave Complete	Price Weight	$4.95 51	$5.45 64	$8.20 91	$12.90 149	$18.00 198
Sheave Only	Price Weight	$2.70 29	$3.30 41	$5.60 57	$9.10 91	$14.05 124
Base and Shaft	Price Weight	$2.60 21	$2.70 22	$3.60 33	$5.65 57	$6.90 73

"CARD" STYLE "B" ROLLER BEARING CURVE SHEAVES

TIMKEN ROLLER BEARINGS

An improved horizontal type curve sheave with bearings so designed that both thrust and radial loads are equally taken care of, and with the combined advantages of ease of running and infrequent lubrication.

They are grease lubricated direct to an ample grease chamber and are dirt-proof, being protected both top and bottom against the entrance of foreign matter.

Regularly equipped with a standard fitting for grease lubrication.

Cap is so designed that rope cannot catch, should it miss the face.

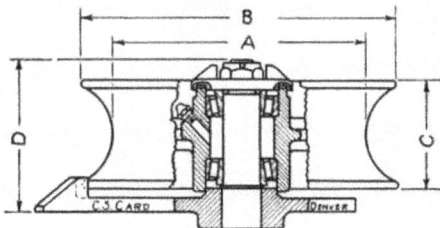

The faces of the sheaves are chilled to minimize rope wear. Sheaves are made of "CARD" semi-steel wheel metal and heat-treated.

On account of longer life for both rope and sheave, we recommend the use of the largest diameter wherever possible.

All sizes carried in stock for immediate shipment.

Specifications; Prices and Average Weights in Pounds, Each

Sheave Number		361	372	384	397
Groove Diameter	"A"	8½"	12"	18"	24"
Outside Diameter	"B"	10½"	15"	21"	27"
Width of Face	"C"	5½"	5"	5¼"	5¼"
Height	"D"	7"	7¼"	7¾"	7¾"
Curve Sheave Complete	Price	$11.60	$14.50	$18.70	$24.30
	Weight	62	90	149	222
Sheave Only, Without Bearings	Price	$ 5.05	$ 7.15	$10.40	$15.95
	Weight	38	54	92	154
Base and Shaft	Price	$ 4.30	$ 4.85	$ 6.30	$ 7.45
	Weight	20	30	51	63

"CARD" PIPE TRACK ROPE ROLLERS

PLAIN BEARING

This is an inexpensive type of track roller. It is made with chilled iron bearings and has a renewable pipe filling. Chilling the bearings adds greatly to their life.

There is a capacious oil reservoir in each box. The standard length of filling is 14" and will be furnished unless otherwise specified, but any length can be furnished.

This roller can also be furnished with oak filler of any length and up to and including 5½" diameter.

Order by roller number and specify length of filling.

Prices listed are for standard 14" length of filling.

All sizes carried in stock for prompt shipment.

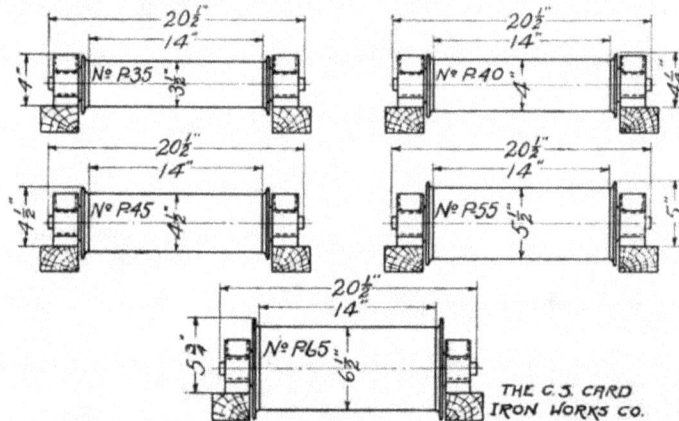

Diameter indicated is the approximate outside diameter of the pipe.

Specifications; Prices and Average Weights in Pounds

Roller Number		P-35	P-40	P-45	P-55	P-65
Roller, Complete with Journal Boxes, each	Price	$3.10	$3.35	$3.65	$4.40	$5.15
	Weight	29	31	34	39	47
Roller, without Boxes, each	Price	$2.25	$2.50	$2.85	$3.55	$4.35
	Weight	17	19	22	27	35
Journal Boxes, per pair	Price	$0.95	$0.95	$0.95	$0.95	$0.95
	Weight	12	12	12	12	12
Roller End Casting, per pair	Price	$0.62	$0.67	$0.72	$0.94	$1.16
	Weight	7	7½	8	8	11
Add or subtract for each inch difference in length from standard 14" filling		$0.06	$0.08	$0.09	$0.11	$0.15

"CARD" ROLLER BEARING TRACK ROPE ROLLERS

SPECIAL ANALYSIS STEEL TUBE—TIMKEN ROLLER BEARINGS

An easy running, well lubricated roller built of special materials to give the maximum life with the least attention.

The ROLLER is in perfect balance on the bearings and turns at the lightest touch of the rope. It is grease lubricated, with a grease chamber large enough that the roller will run at least three months on one greasing. The ends of the roller are cast steel; the shell is of special analysis steel to resist rope wear, having approximately five times the life of black pipe. This makes a light roller, easy to turn, yet strong enough to resist considerable abuse. Equipped with standard grease fittings located so they cannot be struck and broken off.

SHAFT is special steel of our mine car wheel roller bearing specifications. It is not weakened by oil holes, or other machining, and has a liberal bearing in the boxes.

BEARING BOXES are cast iron, into which the shaft is assembled by a cotter. By removing the cotter, the roller, roller bearings and shaft can be removed as a unit without disturbing the boxes. The low Style "D" box will then permit a locomotive or mining machine to pass with a minimum clearance. The boxes are so designed that in case the roller is struck in a wreck or derailment they will let go before serious damage is done to the roller proper or shaft. Style "D" boxes furnished regularly. Special Style "E" boxes with rope guard will be furnished only when specified.

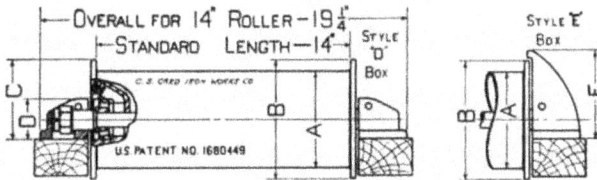

The roller can be made any length. Unless otherwise specified we will furnish with a standard filler 14" long, as shown. In ordering, please specify the roller number, or diameter, and the length of filler, if other than standard 14".

See also "CARD" Roller Bearing Slope Roller, page 74.

Specifications; Prices and Average Weights in Pounds

Roller Number		45 PT		55 PT		66 PT	
Roller Diameter	"A"	$4\frac{1}{2}$"		$5\frac{9}{16}$"		$6\frac{5}{8}$"	
Flange Diameter	"B"	$5\frac{1}{2}$"		$6\frac{11}{16}$"		$7\frac{7}{8}$"	
Flange Height	"C"	$3\frac{5}{8}$"		$4\frac{1}{4}$"		5"	
Box Height	"D"	$1\frac{7}{8}$"		$1\frac{7}{8}$"		$2\frac{1}{4}$"	
Box Height	"E"	4"		$4\frac{5}{8}$"		$5\frac{3}{8}$"	
		Price	Weight	Price	Weight	Price	Weight
Roller Complete with Style "D" Boxes, each		$7.50	26	$9.35	36	$12.75	41
Roller Complete with Style "E" Boxes, each		7.85	30	9.75	42	13.30	47
Roller Only, (Shell and Ends), each		4.80	19	6.90	28	9.05	31
Boxes, per pair (Style "D")		0.70	3	0.70	3	0.90	5
Boxes, per pair (Style "E")		1.05	7	1.10	9	1.45	11
Extra–Each additional inch over 14" standard length filler		0.15	1	0.25	$1\frac{1}{2}$	0.30	2

These rollers are manufactured in large quantities and carried in stock in standard 14" length. Rollers shorter than standard length take standard price list.

"CARD" ROLLER BEARING SLOPE ROLLERS

MANGANESE STEEL SHELL—TIMKEN ROLLER BEARINGS

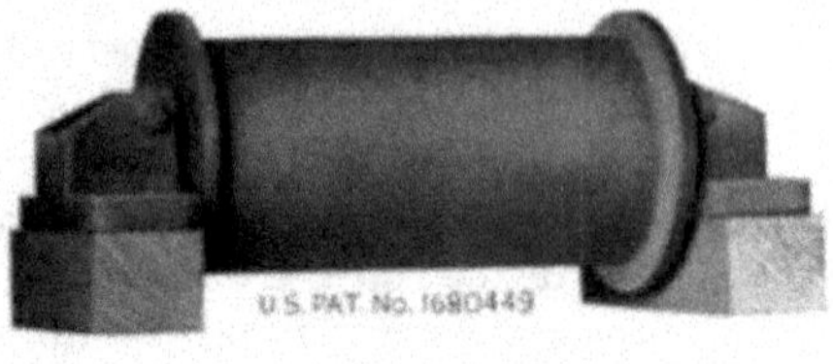

A roller designed especially for heavy duty, large ropes and high speeds.

Operators often use this roller to take the heavier loads on the high spots on a slope or incline and use lighter, less expensive rollers, such as shown on page 73, between these high points. We recommend this practice. Both style rollers have the same grease fittings.

The roller shell is manganese steel, one-half inch thick. The roller is well balanced, both the roller shell and roller ends being balanced before being assembled.

The heavy high carbon steel shaft is not weakened by oil holes or other machining.

The roller, shaft, and roller bearings are assembled as a self-contained unit and cannot bind when set in the bearing boxes. It is very easy to assemble and disassemble these parts when necessary.

The roller is equipped with Timken bearings, well sealed and grease lubricated. It runs at least three months on one lubrication, under average mine operating conditions.

Shipment of roller No. 65-14 MST can be made from stock, grease packed, ready for installing; other sizes promptly.

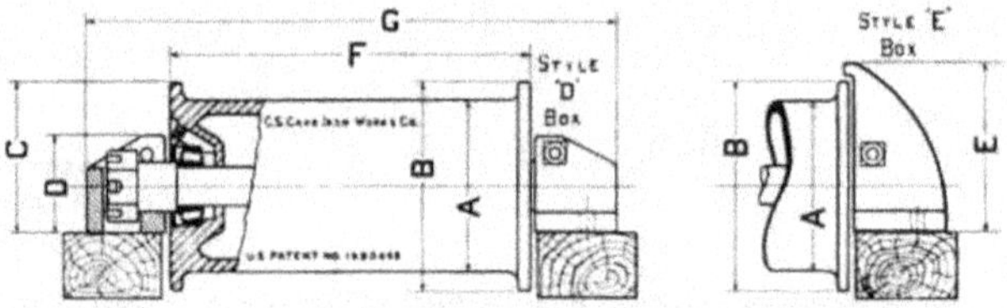

Specifications; Prices and Average Weights in Pounds

Roller Number		65-14 MST		85-12 MST		110-10 MST	
Roller Diameter	"A"	6½"		8½"		11"	
Flange Diameter	"B"	8"		10"		12½"	
Flange Height	"C"	5¾"		6¾"		8"	
Box Height	"D"	3⅝"		3⅝"		3⅝"	
Box Height	"E"	6⅜"		7½"		9"	
Roller Length	"F"	14"		12"		10"	
Overall Length	"G"	20¾"		18¾"		16¾"	
		Price	Weight	Price	Weight	Price	Weight
Roller Complete with Style "D" Boxes, each		$26.50	78	$31.40	95	$34.90	110
Roller Complete with Style "E" Boxes, each		29.00	87	34.60	110	38.90	130
Roller Only, (Shell and Ends), each		20.50	44	26.00	61	30.10	76
Style "D" Boxes, per pair		2.00	18	2.00	18	2.00	18
Style "E" Boxes, per pair		4.50	27	5.20	34	6.00	40

"CARD" CONCAVE CHILLED FACE SLOPE ROLLERS

PLAIN AND ROLLER BEARING

Both the plain and roller bearing rollers are made of "CARD" semi-steel wheel metal, are heat-treated, and have a chilled face to minimize the wear from the rope.

All sizes are carried in stock.

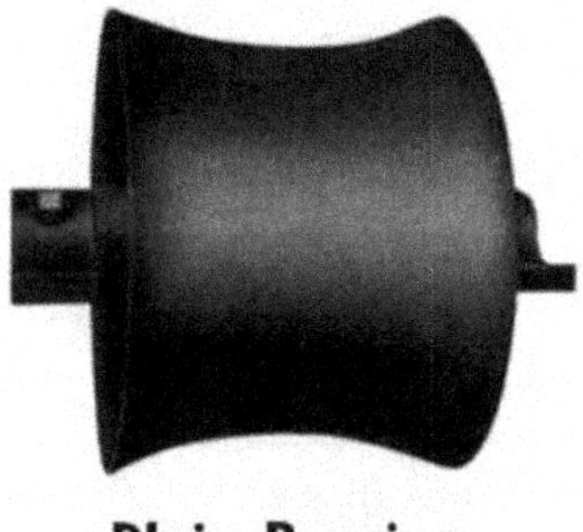

Plain Bearing

The plain bearing rollers run loose on the shaft which is set-screwed in the solid cast iron boxes.

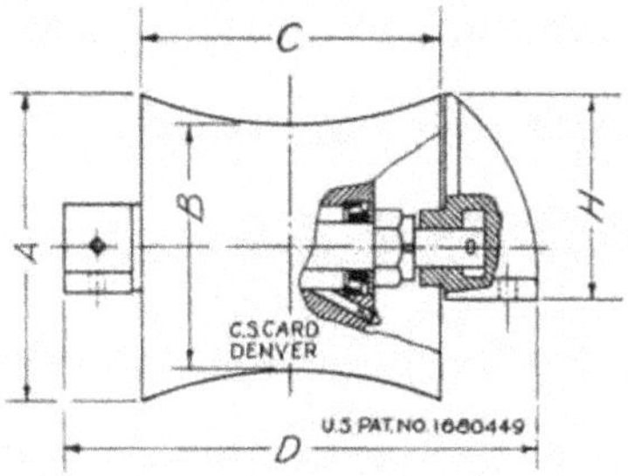

Plain and Roller Bearing

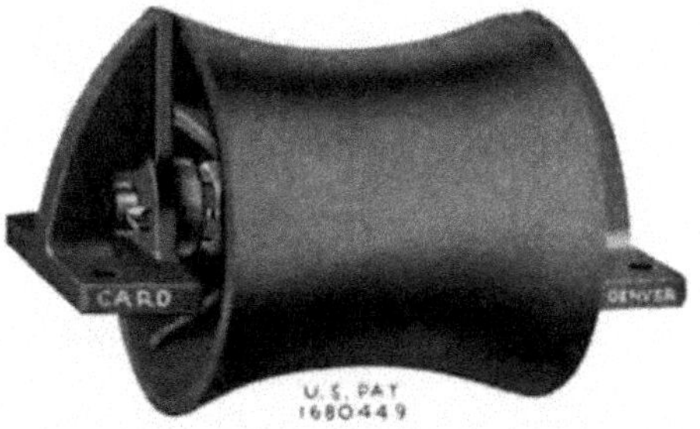

Roller Bearing

Timken roller bearings are used in these rollers, which require little rope friction to start them revolving and keep them turning at rope speed.

The roller cannot be set up wrong so that it binds in the boxes, or so that a rope can get between the roller and boxes.

The sealed bearings are dirt-proof and easily pressure lubricated by grease, packed directly around the bearings in an ample grease chamber.

The boxes have a rope guard to prevent the rope lying on top and sawing and to assist the rope in getting back on the roller.

Specifications; Prices and Average Weights in Pounds, Each

		Plain Bearing						Roller Bearing	
Roller Number		65	69	107	111	115	117	311	315
Outside Diameter	"A"	5"	6"	8"	9"	12"	11½"	9"	11"
Diameter at Center	"B"	4"	4½"	6¾"	7¼"	10¼"	9½"	7¼"	9¼"
Roller Length	"C"	4"	6"	5⅞"	9"	9"	12"	9"	9"
Overall Length	"D"	7¾"	10½"	10⅜"	13½"	13½"	16½"	14¾"	14¾"
Height	"H"	3½"	4 5/16"	5 5/16"	5 13/16"	7 5/16"	7 1/16"	6"	7"
Shaft Size		¾"	1"	1¼"	1¼"	1¼"	1¼"	1⅜"	1⅜"
Roller Complete	Price	$3.80	$5.20	$7.25	$8.60	$9.45	$11.20	$14.15	$15.30
	Weight	17	30	38	51	74	89	59	73
Roller Only	Price	$2.70	$4.05	$5.95	$7.30	$8.20	$10.00	$7.35	$8.45
	Weight	8	20	26	39	62	76	39	51

"CARD" CHILLED GROOVE SLOPE ROLLERS

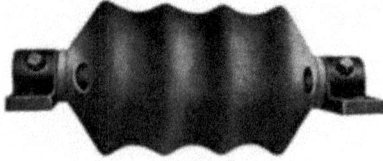

A popular style slope roller that turns on a special analysis high carbon steel shaft held stationary in plain boxes. The balanced roller can turn freely although the boxes may not be in perfect alignment. The sloping ends allow the rope to climb back on the roller more readily than many styles of rollers.

This roller is more popular when equipped with graphite bronze self-lubricating bushings.

All sizes carried in stock for immediate shipment.

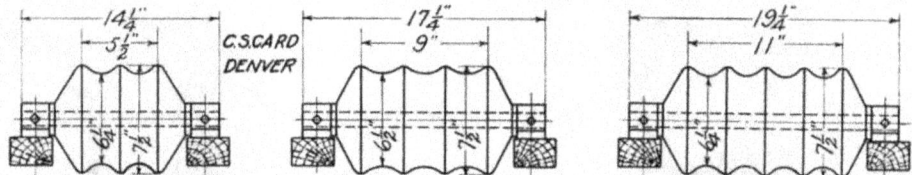

Specifications; Prices and Average Weights in Pounds, Each

Roller Size	Two Groove		Three Groove		Four Groove	
	Price	Weight	Price	Weight	Price	Weight
Roller complete, plain bearing	$5.10	35	$5.70	45	$6.25	53
Roller only, plain bearing	3.95	24	4.50	33	5.10	41
Roller complete graphite bronze bushed	7.40	35	8.00	45	8.55	53
Roller only, graphite bronze bushed	6.55	24	7.10	33	7.65	41

OAK ROLLERS

We are prepared to furnish turned oak rollers of the sizes shown, from No. 1 Car Oak.

Style 1 is a plain oak roller turned to the diameter shown in the table and any specified length.

Style 2 is a turned oak roller with the shaft running through the roller and extending 3″ on each end. Cast iron journal boxes can be furnished, but will only be furnished when specified.

Rollers furnished standard 12″ length unless otherwise specified.

See page 72 of this section for oak rollers with chilled cast iron ends and boxes.

Specifications; Prices and Average Weights in Pounds, Each

		Style No. 1			Style No. 2		
Approximate Outside Diameter		3″	4″	6″	3″	4″	6″
Standard Length of Roller		12″	12″	12″	12″	12″	12″
Shaft Diameter		....	....	....	¾″	¾″	1″
Roller Only	Price	$0.80	$1.05	$1.40	....	....	....
	Weight	3	5	12	....	....	....
Roller and Shaft	Price	....	....	....	$1.55	$1.80	$2.35
	Weight	....	....	....	5	7	16
Add or subtract for each inch difference in length from standard 12″ roller		$0.08	$0.10	$0.14	$0.10	$0.12	$0.17

Boxes for Style No. 2 rollers: $1.90 per pair; weight 10 pounds per pair.

"CARD" STYLE "F" ROLLER BEARING CURVE SHEAVES
TIMKEN ROLLER BEARING

Similar in design to our Style "B" sheaves, shown on page 71, Section "D".

Style "F" sheaves turn easily and quickly reach the rope speed with little slippage between the sheave and rope.

They are grease lubricated direct to an ample grease chamber and are designed so the lubricant stays in and dirt stays out of the bearings. Both cost and frequency of lubrication are reduced to a minimum.

Sheaves are made of our semi-steel wheel metal and heat-treated. The chilled groove minimizes rope wear.

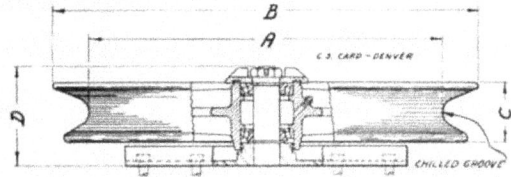

Specifications; Prices and Average Weights
in Pounds, Each

Sheave Number		30-F	36-F
Groove Diameter	"A"	30"	36"
Outside Diameter	"B"	36"	42"
Width of Face	"C"	4⅞"	4⅞"
Height	"D"	8¼"	8½"
Curve Sheave Complete	Price	$42.30	$54.30
	Weight	375	445
Sheave Only	Price	$29.50	$41.45
(without bearings)	Weight	315	380
Base and Shaft	Price	$11.20	$12.05
	Weight	50	55

"CARD" CARRYING SHEAVES WITH CLEVIS

These sheaves are designed for carrying and guiding ropes. The sheaves run idle on a pipe spacer, which can be easily renewed when worn. Designed so the rope cannot get between the sheave and clevis.

Sheaves No. 214, No. 121, and No. 381 have chilled grooves. The groove of sheave No. 214 is large enough to pass a rope coupling.

Sheaves can be furnished with self-lubricating bushings when used on towers or other places where it is difficult to lubricate them. Made to order. Prices on application.

Eyebolts are not furnished unless specially ordered as an extra. If required, specify the bolt diameter and length under the eye.

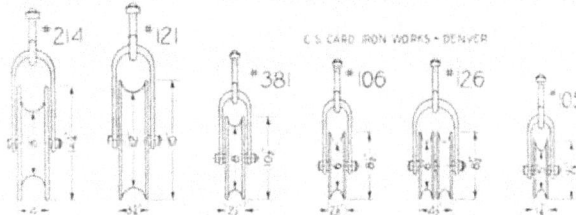

Specifications; Prices and Average Weights in Pounds, Each

Sheave Number		105 Single	106 Single	126 Double	214	121	381
Sheave complete, without eyebolt	Price	$4.50	$4.80	$6.65	$9.60	$8.00	$6.10
	Weight	18	24	39	73	52	30
Sheave only	Price	$1.65	$1.80	$1.80	$6.85	$4.85	$3.15
	Weight	9	15	15	63	42	24

"CARD" SWIVEL, WITH SOCKET

Our Swivels and Sockets are made extra heavy throughout, and are forged from special steel. The pin is so constructed as to be practically as strong as though both heads were forged solid, without a joint. Sizes listed below are standard stock sizes and include one closed rope socket.

Special sizes made to customers' specifications on short notice.

All listed sizes carried in stock for prompt shipment.

Specifications; Prices and Average Weights in Pounds, Each

Swivel Number	00	0	1	2	3
Size of Rope	1½″	1⅜″	1¼″	1″	⅞″ & ¾″
Size of Pin	1¾″	1½″	1⅜″	1¼″	1⅛″
Price	$46.00	$40.00	$32.30	$27.25	$25.25
Weight	54	43	38	26	21

"CARD" ROPE HITCHINGS, WITH SWIVEL

Rope Hitchings are made of best grades of special steel throughout, carefully forged and designed to meet local conditions. Made to order promptly upon receipt of specifications.

"CARD" ROPE SAFETY CLAMP

A safety device for use with a rope connection to cars, man trips, skips or cages, used on inclines or shafts. It protects against failure of the rope at the socket or of the connection between the rope socket and cars, etc. In case of such a failure, the load is transferred through the safety chains to the clamp and rope ahead of socket.

The clamp is of alloy steel, and heat-treated to develop maximum strength.

The clamp and safety chains can be easily removed from the conical safety button which is made a permanent part of the rope.

Blue print describing installation and giving additional details upon request.

In ordering, specify wire rope diameter and size of center core in the rope. Made to order upon receipt of specifications.

Specifications; Prices and Average Weights in Pounds, Each

Clamp Number	1	2	3
Rope Size	1½″ to 1¾″	1⅛″ to 1⅜″	⅝″ to 1″
Price	$48.00	$44.00	$40.50
Weight	29	24	20

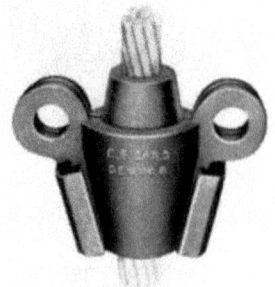

A

SECTION "E"
CATALOG NO. 40

TIPPLE EQUIPMENT
DUMPS, CAGES, SCREENS, Etc.

The C.S. Card Iron Works Co.
Denver, Colorado

"CARD" TIPPLE EQUIPMENT

SCREENS, DUMPS, CAGES, ETC.

Since 1892 we have been designing and manufacturing equipment for coal tipples and are thoroughly familiar with the screening requirements of the Rocky Mountain Coal Fields. No two plants have been duplicates. Our engineering department is prepared to submit designs and specifications on equipment for your own special requirements.

On the following pages we show but a few of the coal screening plants we have designed and furnished the equipment for.

During this period we have had a wide experience in developing different designs of car dumps, cages, coal screens, etc. Cages, dumps and skips are illustrated in a few of the different designs we have furnished. Since this equipment is designed and built to meet operating requirements, it is impractical to show all the different designs with their many variations.

This modern steel tipple with bins and headframe was designed and furnished erected for
The Imperial Coal Company, Erie, Colo.

The Leyden Lignite Co., No. 3 Mine, Leyden, Colo.

Union Pacific Coal Co., No. 4 Mine, Rock Springs, Wyo.

William E. Russell Coal Co., Frederick, Colo.

Rock Springs Fuel Co., Superior, Wyo.

Butte Valley Coal Co., Alamo, Colo.

McNeil Coal Corporation, Sterling Mine, Dacona, Colo.

"CARD" LOADING BOOMS

"CARD" loading booms are built with either steel aprons or with rubber belts for carrying the coal.

The width and speed of the apron is designed according to whether or not picking of the impurities from the coal is desired and the capacity required.

The illustration is of a steel apron loading boom, with the complete drive and the pickers' platform attached to the steel frame, making an integral unit. The boom is raised and lowered or is swung in an arc, as a unit. This type boom is used for loading direct into open cars or is swung parallel with the track along the side of a box car into which the coal is transferred by a chute or box car loader.

Loading booms are built on order to meet the operating requirements. Information, prices, etc., furnished on receipt of your requirements.

"CARD" AUTOMATIC OR SELF-DUMPING CAGES

Short Radius Dump Type

We illustrate two types of automatic or self-dumping cages in which the platform with the car on it is dumped by means of dump rollers that engage in dump shoes.

The short radius dump type cage dumps quickly on engaging the dump shoes, while the long radius dump type action is slower and less severe.

Cars, with lift doors, are locked on the cage by various mechanisms while the cage is being raised and lowered. The locking mechanism is usually unlocked automatically at the shaft bottom, so the car can be bumped or run off the cage. A manually operated device for unlocking is provided for use at the ground landing. Spring cushioned horns for stopping the oncoming loaded car on the cage are frequently used.

Various mechanisms prevent the cage from dumping at any point in the shaft except at the dump shoes.

Safety dogs, that grip the shaft guides in case of accident, are built in several different styles. The cages shown have heavy steel safety dogs that are positive in action, operated by both springs and gravity.

Our self-dumping cages are substantially constructed of steel. They are designed and built on order to meet the operating requirements.

Shoes for dumping these cages are supplied when specified.

In ordering or inquiring for information, please furnish us the shaft dimensions as per outline drawing on page 86. Also furnish the total gross load of car and material to be handled, and a drawing of the car that will be used.

Long Radius Dump Type

"CARD" PLAIN CAGES

These three plain cages illustrate only a few different designs we have furnished. They serve to show different types of details such as headframe and safety dogs, sides, doors, platforms, etc.

Plain cages are built to meet the operating conditions. In ordering or inquiring for information, please furnish us the shaft dimensions, as per outline drawing on page 86, and the total gross load to be handled. Advise whether or not doors are required.

See page 86 for landing chairs for use with plain cages.

Cage No. 2011

Cage No. 2622

Cage No. 1876, with landing chairs built in the cage

"CARD" LANDING CHAIRS

The chairs are made of mild steel throughout, excepting bearings.

The chairs are set by the lever and the supports are pulled into the clear by the springs when the cage is raised.

The opposite operation can be obtained by placing the spring below the pivot shaft.

A print showing the setting of these chairs will be furnished on request.

In ordering, please furnish the information requested below, as these chairs are built on order for the shaft size.

Price, each complete set $80.00

Approximate weight, each complete set .450 lbs.

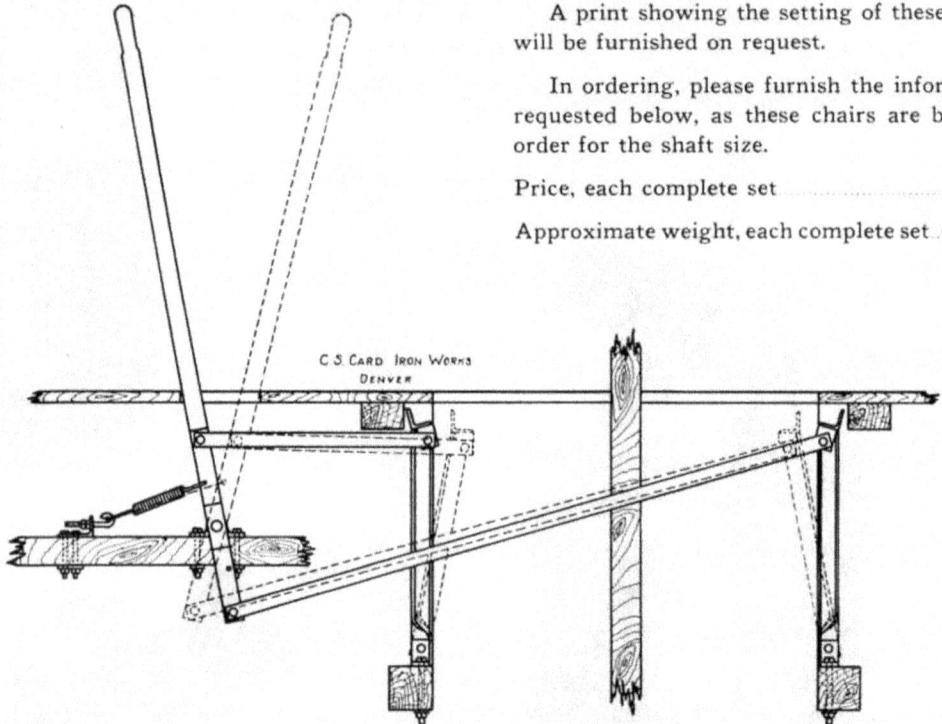

See Cage No. 1876, Page 85, Section "E", for automatic landing chairs built into the cage.

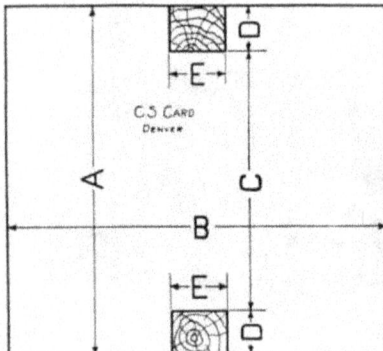

Shaft Dimensions

This sketch shows the shaft dimensions required by us to correctly furnish cages, landing chairs, skips, etc.

"E" to be the actual width of guide as we will allow the necessary clearance.

"A" and "B" only, required for plain landing chairs.

"CARD" SKIPS

Skips for hoisting in vertical shafts are designed and built to meet operating conditions. It is impractical to illustrate all the variations used in skip design for this purpose.

The two self-dumping skips shown have different styles of safety dogs for gripping the shaft guides in case of accident. Skip No. 1880 has forged steel dogs, operated by both springs and gravity, that are forced outward into the wood guides. Skip No. 2654 has dogs with teeth that grip the guides from the sides. These dogs are mounted on a through shaft and are operated by springs.

Dump shoes for self-dumping skips can be supplied when specified.

In ordering or inquiring for information on these skips, please furnish us the shaft dimensions as per outline drawing on page 86. Also advise the body capacity desired in cubic feet, and weight of the material per cubic foot that will be handled.

Skip No. 2654

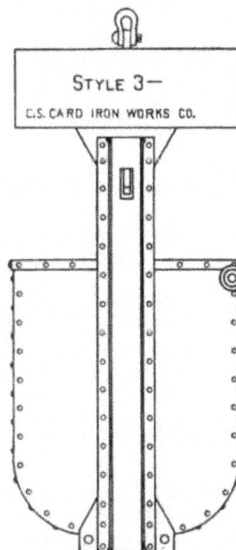

We designate skips as shown in the outline drawing as our Style No. 3. See page 62, Section "C", for Styles No. 1 and No. 2 Skip Cars used on inclines.

Skip No. 1880

"CARD" ROTARY DUMPS

POWER DRIVEN

Potash Company of America, Carlsbad, New Mex.

We illustrate but a few of the rotary dumps we have furnished. We build several different types of power driven rotary dumps. In all instances the dumps have been designed especially to meet the requirements peculiar to each condition encountered.

The dump illustrated above was designed for handling a train of cars without uncoupling, swivel couplings being used and the cars being dumped one at a time. The dump was built in sections that could be lowered through the shaft and was assembled underground.

The principal advantage in a rotary dump is that it permits the use of a car with solid ends and sides, the most substantial car design there is. The initial investment in the dump is offset by a saving in the first cost of the cars, lower upkeep and fewer repairs to the car, and a tight car body with consequently less spillage along the haulage ways.

In handling coal, with a properly designed dump chute, the breakage is reduced to a minimum.

Cars are ordinarily dumped at the rate of two to three cars per minute.

Inasmuch as each dump installation has its own peculiar problems, we request that you furnish us as full a picture of your requirements as possible when inquiring for dump information. We will be glad to furnish specifications and prices upon receipt of your requirements.

"CARD" ROTARY DUMPS

POWER DRIVEN

In the dump illustrated above the simple throwing of one lever forward and then back causes the dump to revolve and sets the stop so the dump automatically stops in the correct position to receive the next car. In dumping a train, it is unnecessary to start and stop the dump motor with each car dumped.

The dump shown at the side was furnished complete with the frame on which it is mounted. The rotary dump proper is carried on four wheels mounted on through shafts that are carried in extra heavy roller bearing self-aligning journal boxes. Pushing a button starts the dump revolving and it is stopped in the correct position by a brake on the motor and lock on the dump.

"CARD" ROTARY DUMPS
GRAVITY OPERATED

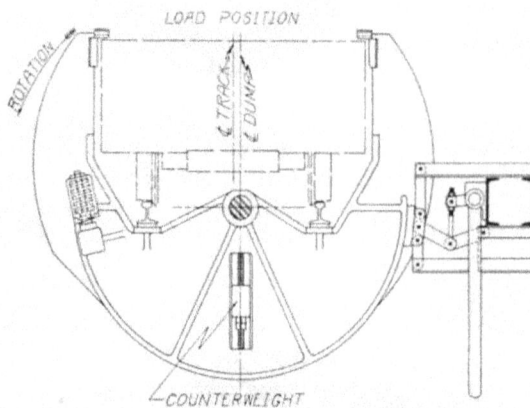

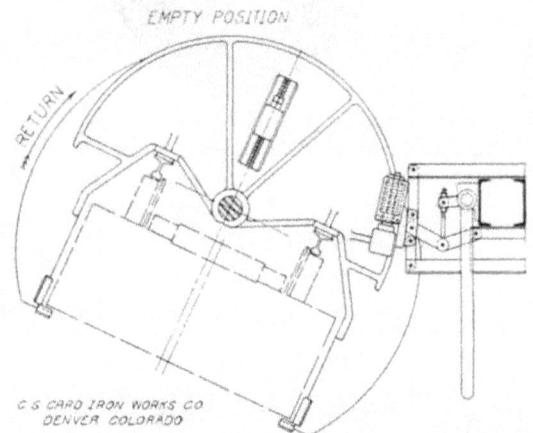

A dump which will operate where no power is available. The operation of the dump is entirely by gravity. Both installation and operation costs are low. The simplicity of design and few operating parts alone make it popular. To this must be added the advantages of the solid end cars used.

The dump can be locked in any position by the brake which operates on the machined surfaces on the rings. The rotating speed is controlled by this brake.

In the usual operation, the loaded car is locked on the dump by retaining angles, horns, etc. The brake is released and car and dump rotate through an arc to a position where the material will pass from the car. Counterweight in the dump returns it and the empty car to normal position and the cycle is complete. Spring cushioned stops limit the travel in both directions. A loaded car run on to the dump is usually used to bump the empty off.

Three cars per minute can be dumped easily and under the right conditions five cars.

Cars must be uncoupled and dumped one at a time with this type dump.

Specifications will be furnished and prices quoted upon receipt of your requirements.

"CARD" GOOSE-NECK DUMP
WITH SPRING JOURNAL BOXES

A popular, inexpensive type of dump, which operates by gravity, for coal tipples, rock dumps, etc.

It is substantially constructed, of steel throughout, and will withstand the most severe service. It is equipped with the spring journal boxes listed below. We recommend the spring journal boxes as they save unnecessary punishment and consequent repairs to the dump, car and tipple.

In ordering, specify the track gauge, diameter of the wheels and the wheel base (distance center to center of axles). If the wheel base is off the center of the car, send sketch showing the location.

Built in two weights. We will gladly specify the one best suited for your service if advised the gross weight of car and material combined, which will be used.

	Standard Dump Number 201			Extra Heavy Dump Number 1155		
Track Gauge	30″	36″	42″	30″	36″	42″
Price, Each complete with Spring Boxes	$102.50	$105.00	$107.50	$115.00	$120.00	$125.00
*Weight, Pounds, Each with Spring Boxes	475	510	540	670	700	750

*The weights listed are for average sizes. They vary according to wheel base and wheel size.

"CARD" TIPPLE GOOSE-NECK SPRING JOURNAL BOXES

The "CARD" Tipple Goose-Neck Spring Journal Box is a great saver of shock, wear and tear on the car and tipple; it saves breakage of car axles; it prevents, by taking up the shock, the loosening of the axle on the car-body, and saves unnecessary pounding on the tipple frame. It can be made to fit any Goose-Neck.

In ordering, specify the size the box is to be bored. Made in two sizes, Standard for bearings up to 2″ diameter; Extra Heavy for bearings up to 2½″ diameter.

PRICE, per pair, Standard Pattern Number 201 .... $35.00

Weight, per pair, 130 pounds.

PRICE, per pair, Extra Heavy Pattern Number 1155 .... $40.00

Weight, per pair, 220 pounds.

"CARD" CROSSOVER DUMP

This popular type of dump is operated by gravity. The dump is held in normal position by a brake. A loaded car is run on the dump. The brake is released, allowing the car to dump and unload at an angle of about forty-five degrees. When unloaded the car and dump are returned to normal position by counterweights. Another loaded car running on to the dump opens the horns and bumps the empty car off the dump. The loaded car is caught by the horns and the cycle of operation is complete.

Cars with end doors that can be swung or raised are used with this dump.

The dump is substantially constructed entirely of steel with the exception of the weights and rockers. It can be built to handle any size car.

In ordering, specify the gross weight of the car and the load combined, the track gauge, wheel base, diameter of the wheels, and the overall length of the car.

We build this dump in two weights. We will furnish the weight best suited if advised the combined weight of car and material the dump is to handle.

	Standard Dump Number 1062			Extra Heavy Dump Number 1061		
Track Gauge	30″	36″	42″	30″	36″	42″
Price, Each complete	$485.00	$505.00	$515.00	$525.00	$540.00	$555.00
*Weight, Pounds, Each complete	2700	2900	3100	3000	3250	3500

*The weights listed are for average sizes. They vary according to wheel base, wheel size, length of car overall, etc.

"CARD" REVOLVING SCREENS

The illustration shows but one of many styles of Revolving Screens we have made for the different industries.

We make Revolving Screens with both single and double jackets, any specified length or diameter, covered with either perforated plate or double crimped wire cloth, with openings to meet your requirements. Our screens are designed so the jacket can be readily changed.

Send us your specifications, or we will furnish prices and specifications upon receipt of your requirements.

DUO-PURPOSE

TRUCKLOADER

SCREENS AND LOADS

DUO-PURPOSE

"CARD" TRUCKLOADER

SCREENS AND LOADS

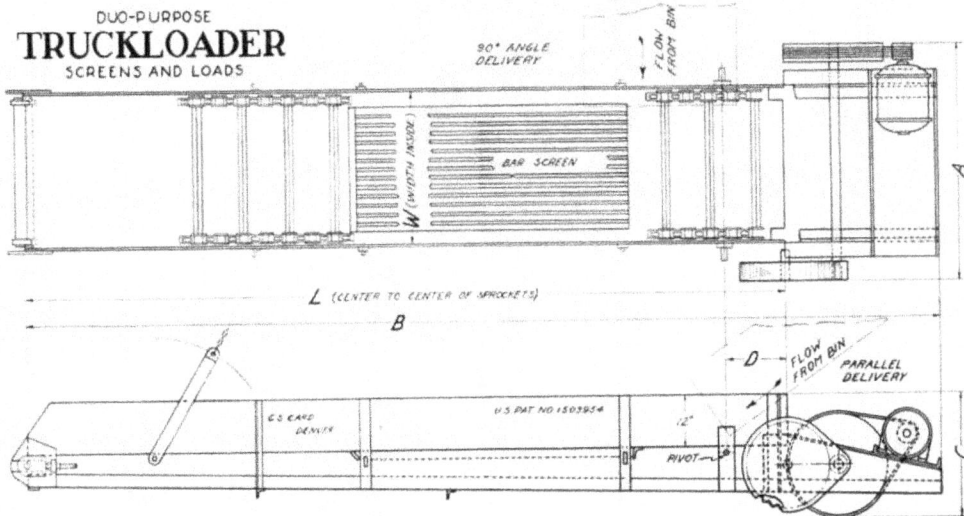

The TRUCKLOADER fills the demand for a suitable loading and screening device for handling sized coal from bins to trucks, with a minimum of breakage and with the coal thoroughly screened.

It is a self-contained unit, easily installed and readily supported from a bin. The delivery end is raised or lowered to suit loading conditions.

The coal is carried over screen bars, the undersize falling to the bottom plate where it is picked up by the return strand of the conveyor and delivered at the rear end of the TRUCKLOADER. By adjusting the openings between screen bars as much or as little of the undersize can be removed as is desired.

The TRUCKLOADER will screen and load coal to trucks at a rate of one to three tons per minute.

In ordering, specify the size by number. Specify voltage for motor. Advise size of coal to be handled, size of screen bar opening desired, and whether coal will be delivered parallel with or at an angle to the TRUCKLOADER. Parallel delivery is recommended.

Specifications; Prices and Average Weights in Pounds, Each

Number	W	L	A	B	C	D	Price	Weight
TL–1	36"	15'–0"	4'–9"	18'–0"	30"	15" or 36"	$585.00	2700
TL–2	36"	12'–0"	4'–9"	15'–0"	30"	15" or 36"	$560.00	2450
TL–3	28"	15'–0"	4'–1"	18'–0"	30"	15" or 36"	$525.00	2500
TL–4	28"	12'–0"	4'–1"	15'–0"	30"	15" or 36"	$500.00	2300

Prices and weights cover complete standard unit as illustrated, with 3 H.P. ball bearing motor, 3 phase, 60 cycle, either 220 or 440 volts; with drive. No starter or wiring is included. Prices on special sizes on request.

Standard "D" is 15" for parallel delivery; 36" for angle delivery to TRUCKLOADER.

Above drawing shows right hand assembly of motor, with drive. Opposite or left hand assembly can be made when so specified.

"CARD" PUNCHED STEEL SCREEN PLATES

We are prepared to furnish you promptly with perforated plates, punched to your order.

In ordering, give us the following information:

The width, length and thickness of the plate.

The size perforations and shape of the holes.

Width of margin required, if special.

Unless special centers between perforations are ordered, we will furnish our standard.

Specify with which dimension the material flows so the perforations can be properly staggered.

If bolt holes are required, furnish dimensioned sketch showing size and location of holes.

If plates of special shape are desired, furnish us with a sketch showing the dimensions.

Plates rolled to fit a cylinder, or a section of a cylinder can be furnished upon receipt of specifications. Complete cylinders, with or without butt straps, can be furnished.

"CARD" BAR OR GRAVITY SCREENS

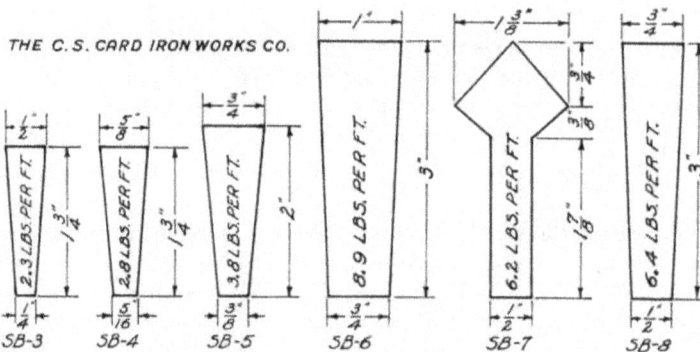

We have made numerous installations of Gravity Screens.

We are prepared to furnish the bars and rests only or the complete chute with bars and rests. Any style bar with any size opening can be furnished. The different sizes of bars most commonly used are illustrated.

In asking for prices, furnish the following information:

The section number of the bar and the length.

The width of the opening between the bars.

The inside width of the chute or overall length of the screen bar rest.

Quotations will be furnished upon receipt of this information.

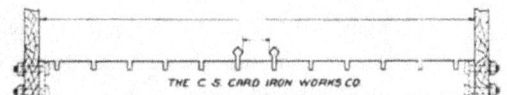

SECTION "F"
CATALOG NO. 40

TRACK EQUIPMENT
FROGS, SWITCHES, CROSSINGS, Etc.

The C.S. Card Iron Works Co.

Denver, Colorado

"CARD" TRACK EQUIPMENT

With the use of heavier locomotives, cars, mining machines and loaders continually increasing, and a greater demand for speed all along the line, better track and track equipment is a necessity. Safe, efficient, economical production demands it.

Low haulage cost is built on the foundation of:—correct weight of rail for the rolling stock used; good ties, properly spaced; solid road-bed, well drained; easy curves and grades; proper track alignment; and the use of suitable track equipment such as frogs, switches, stands, etc.

Substantially built standard track equipment pays for itself in a short time by decreasing wrecks and derailments. One bad wreck will pay for a lot of standard track equipment. Standardization of your track equipment will pay you big dividends. Your track layers become familiar with standard equipment and the laying is speeded up.

American Standards Association Standards

In the following pages the trackwork follows, with minor exceptions, the standards as approved by the American Standards Association. These standards were sponsored by and are also known as American Mining Congress Standards.

Instructions for Ordering Track Equipment

In ordering track material or asking for quotations, always furnish us the following data:

RAIL SECTION: Specify the weight of rail (per yard). If a special section of rail is desired, other than shown on page 118, send a small section of the rail, or a sketch, or a template which can be obtained by placing heavy paper or cardboard against the end of the rail and tapping lightly with a hammer around the edges.

HOLES FOR FISH PLATES OR ANGLE BARS: Furnish us a sketch showing size of holes, distance from end of rail to center of first hole, and distance center to center of holes, if other than standard shown on page 118.

BOND HOLES: Give diameter of hole and distance from the end of the rail to the center of the hole.

FROGS: Give the frog number, or the frog angle in degrees and minutes, or the spread of the frog in inches; and also the frog heel and toe lengths if important. See pages 97 and 108.

SPLIT SWITCHES: Furnish the length of the switch point, gauge of the track; and also the throw if it is important.

SWITCH STANDS: Specify the style switch stand desired and the throw if other than the standard specified. See pages 113 to 115.

On all orders for rail work up to and including 60 lb. rail, we will furnish the standard rail sections and drilling as shown on page 118, unless special rail and drilling is ordered.

FROG DATA

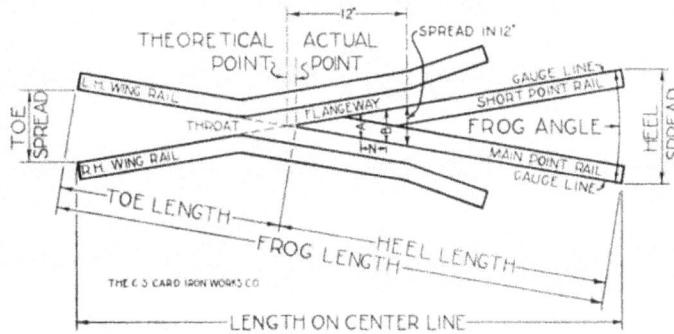

Frog nomenclature

Frog Number

Frogs are designated by number. The number of a frog is determined by the degree of its spread—the angle at which the gauge lines cross. The number of inches in which the spread of the gauge lines increases one inch is assigned as the number of the frog.

To Find the Number of Any Frog: Measure in inches the length of the frog on the centerline and divide by the sum of the heel spread and toe spread in inches (being sure to measure spreads on gauge lines).

Example: A frog measures 60″ long on the centerline. The heel spread is 8″ and the toe spread is 4″. The frog centerline length 60″ divided by the sum of the two spreads 12″ gives the frog number, or a No. 5 frog.

The following is another method also used to determine the number of any frog. Measure across the frog point at any place (as at A) where the distance between the gauge line is an even number of inches; measure again where the distance is an inch greater (B) than at (A). The number of inches (N) between the two measured lines (A) and (B) is the frog number.

Example: (A) measures 3″; (B) measures 4″ (one inch greater); and (N) measures 2½″. It is a No. 2½ frog.

Turnouts are designated by the frog number used; a No. 3 turnout uses a No. 3 frog.

Simple formulas for eliminating guess work in selecting the right frog are given on pages 98 and 99. For your convenience, the tables cover the frog numbers and gauges most commonly used.

Data on standard turnouts are given on pages 100 to 105. The difference between the actual lead on pages 100 to 105 and the theoretical lead in the table on page 98 can be ignored. It is accounted for because the actual lead varies according to the straight frog toe length, thickness of the frog point, length and angle of the switch, and the thickness at switch point.

See page 108 for "CARD" Standard Frog Specifications.

TURNOUT DATA

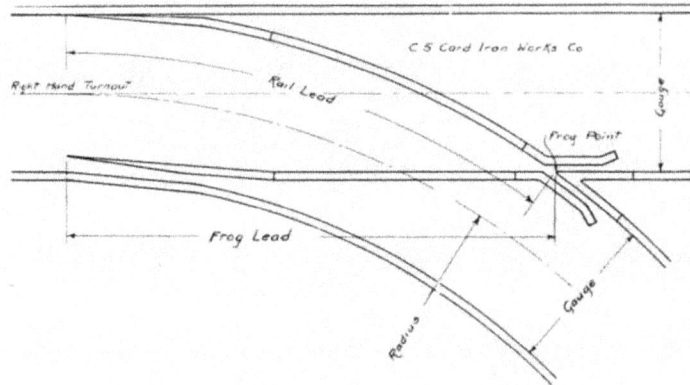

Theoretical Frog Lead of Turnout

The frog lead is the distance from switch point to frog point, measured along the straight rail.

Theoretically it is equal to twice the track gauge in feet multiplied by the frog number.

Example: Track gauge is 36", the frog is a number 4,
then frog lead $= 2 \times 3'\text{-}0'' \times 4$ or $24'\text{-}0''$ lead;

Or reversing the formula:

Example: Frog lead is 24'-0", the track gauge is 36",

$$\text{then frog number} = \frac{24'\text{-}0''}{3'\text{-}0'' \times 2} \text{ or a No. 4 frog.}$$

THEORETICAL FROG LEADS

Frog Number	Track Gauge									
	18"	20"	22"	24"	30"	36"	40"	42"	48"	56½"
1¼	3'-9"	4'-2"	4'-7"	5'-0"	6'-3"	7'-6"	8'-4"	8'-9"	10'-0"	11'-9¼"
1½	4'-6"	5'-0"	5'-6"	6'-0"	7'-6"	9'-0"	10'-0"	10'-6"	12'-0"	14'-1½"
1¾	5'-3"	5'-10"	6'-5"	7'-0"	8'-9"	10'-6"	11'-8"	12'-3"	14'-0"	16'-5¾"
2	6'-0"	6'-8"	7'-4"	8'-0"	10'-0"	12'-0"	13'-4"	14'-0"	16'-0"	18'-10"
2¼	6'-9"	7'-6"	8'-3"	9'-0"	11'-3"	13'-6"	15'-0"	15'-9"	18'-0"	21'-2¼"
2½	7'-6"	8'-4"	9'-2"	10'-0"	12'-6"	15'-0"	16'-8"	17'-6"	20'-0"	23'-6½"
2¾	8'-3"	9'-2"	10'-1"	11'-0"	13'-9"	16'-6"	18'-4"	19'-3"	22'-0"	25'-10¾"
3	9'-0"	10'-0"	11'-0"	12'-0"	15'-0"	18'-0"	20'-0"	21'-0"	24'-0"	28'-3"
3½	10'-6"	11'-8"	12'-10"	14'-0"	17'-6"	21'-0"	23'-4"	24'-6"	28'-0"	32'-11½"
4	12'-0"	13'-4"	14'-8"	16'-0"	20'-0"	24'-0"	26'-8"	28'-0"	32'-0"	37'-8"
4½	13'-6"	15'-0"	16'-6"	18'-0"	22'-6"	27'-0"	30'-0"	31'-6"	36'-0"	42'-4½"
5	15'-0"	16'-8"	18'-4"	20'-9"	25'-0"	30'-0"	33'-4"	35'-0"	40'-0"	47'-1"
5½	16'-6"	18'-4"	20'-2"	22'-0"	27'-6"	33'-0"	36'-8"	38'-6"	44'-0"	51'-9½"
6	18'-0"	20'-0"	22'-0"	24'-0"	30'-0"	36'-0"	40'-0"	42'-0"	48'-0"	56'-6"
7	21'-0"	23'-4"	25'-8"	28'-0"	35'-0"	42'-0"	46'-8"	49'-0"	56'-0"	65'-11"
8	24'-0"	26'-8"	29'-4"	32'-0"	40'-0"	48'-0"	53'-4"	56'-0"	64'-0"	75'-4"

Note: The Actual Lead figures slightly less in most Standard Combinations of Frogs and Switch Points.

TURNOUT DATA

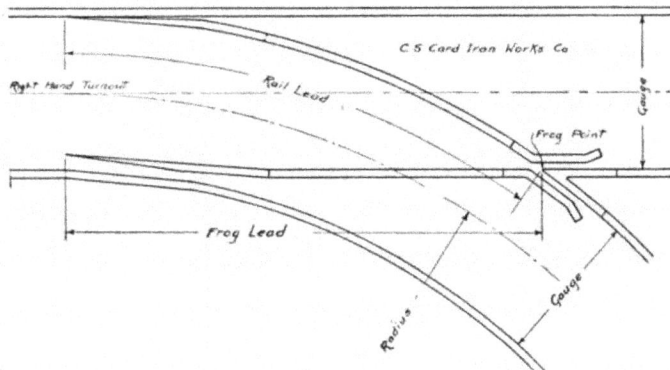

Theoretical Radius of Turnout

The radius or curvature, measured at the center line of the track, is equal to twice the track gauge in feet multiplied by the square of the frog number.

Example: The track gauge is 36″; the frog is a number 4, then radius = 2 x 3′-0″ x 16 or 96′-0″ radius:

Or reversing the formula:

Example: The radius is 96′-0″; the track gauge is 36″,

$$\text{then frog number} = \sqrt{\frac{96'\text{-}0''}{2 \times 3'\text{-}0''}} \text{ or a No. 4 frog.}$$

THEORETICAL RADIUS OF CURVE AT CENTER OF TRACK

Frog Number	Track Gauge									
	18″	20″	22″	24″	30″	36″	40″	42″	48″	56½″
1¼	4′-8¼″	5′-2½″	5′-8¾″	6′-3″	7′-9¾″	9′-4½″	10′-5″	10′-11¼″	12′-6″	14′-8½″
1½	6′-9″	7′-6″	8′-3″	9′-0″	11′-3″	13′-6″	15′-0″	15′-9″	18′-0″	21′-2¼″
1¾	9′-2¼″	10′-2½″	11′-2¾″	12′-3″	15′-3¾″	18′-4½″	20′-5″	21′-5¼″	24′-6″	28′-10″
2	12′-0″	13′-4″	14′-8″	16′-0″	20′-0″	24′-0″	26′-8″	28′-0″	32′-0″	37′-8″
2¼	15′-2¼″	16′-10½″	18′-6¾″	20′-3″	25′-3¾″	30′-4½″	33′-9″	35′-5¼″	40′-6″	47′-8″
2½	18′-9″	20′-10″	22′-11″	25′-0″	31′-3″	37′-6″	41′-8″	43′-9″	50′-0″	58′-10¼″
2¾	22′-8¼″	25′-2½″	27′-8¾″	30′-3″	37′-9¾″	45′-4½″	50′-5″	52′-11¼″	60′-6″	71′-2½″
3	27′-0″	30′-0″	33′-0″	36′-0″	45′-0″	54′-0″	60′-0″	63′-0″	72′-0″	84′-9″
3½	36′-9″	40′-10″	44′-11″	49′-0″	61′-3″	73′-6″	81′-8″	85′-9″	98′-0″	115′-4¼″
4	48′-0″	53′-4″	58′-8″	64′-0″	80′-0″	96′-0″	106′-8″	112′-0″	128′-0″	150′-8″
4½	60′-9″	67′-6″	74′-3″	81′-0″	101′-3″	121′-6″	135′-0″	141′-9″	162′-0″	190′-8¼″
5	75′-0″	83′-4″	91′-8″	100′-0″	125′-0″	150′-0″	166′-8″	175′-0″	200′-0″	235′-5″
5½	90′-9″	100′-10″	110′-11″	121′-0″	151′-3″	181′-6″	201′-8″	211′-9″	242′-0″	284′-10¼″
6	108′-0″	120′-0″	132′-0″	144′-0″	180′-0″	216′-0″	240′-0″	252′-0″	288′-0″	339′-0″
7	147′-0″	163′-4″	179′-8″	196′-0″	245′-0″	294′-0″	326′-8″	343′-0″	392′-0″	461′-5″
8	192′-0″	213′-4″	234′-8″	256′-0″	320′-0″	384′-0″	426′-8″	448′-0″	512′-0″	602′-8″

"CARD" STANDARD TURNOUT DATA

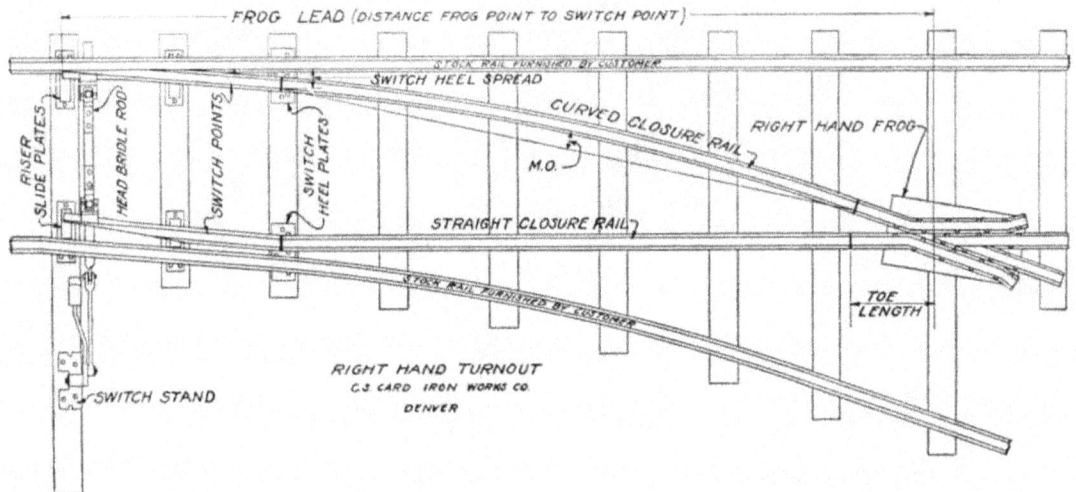

Illustrated is the standard type of turnout which is ordinarily used, with straight frogs and switch points and with the closure rails tangent to the frog and switch points.

For short radius turnouts, particularly in the narrower gauges, we suggest the use of a curved frog turnout with the switch points abutting the frog wing rails. See page 106.

A complete turnout consists of: one frog; one pair of switch points, bridle rods, riser and heel plates, closure rails (one straight, one curved), and one switch stand. Closure rails are sometimes called lead or filler rails.

The trackwork in the following tables follows, with minor exceptions, the standards sponsored by the American Mining Congress and adopted by the American Standards Association.

The following tables, pages 101 to 105, give the actual turnout data covering our standard frogs and split switches. For additional detailed information, see page 108 for frogs and page 110 for split switches.

The following notes apply to these turnout tables:

1. The Actual Lead is based on the Curved Closure Rail being tangent to both the Frog and the Switch Point.

2. The Actual Leads and Radii Curved Rail listed in the following tables are the nearest quarter inch to the mathematical dimensions.

3. The Radius Curved Rail dimension is the Gauge Line Radius.

4. Data for turnouts with other combinations of Frogs and Switch Points can be furnished on request.

"CARD" STANDARD TURNOUT DATA

STANDARD FROGS AND SWITCH POINTS WITH CLOSURE RAILS

On orders for these standard turnouts complete we will furnish one frog, one pair of switch points with clips, necessary bridle rods, necessary riser and heel plates, one pair of closure rails (one straight, one curved), and one ground throw switch stand with plain connecting rod as shown on page 113.

Switch stand may be omitted or other styles substituted, if desired.

Any parts of the turnout may be omitted. Closure rails often are, especially when the turnouts have long leads.

We recommend that you use a turnout with curved frog and switch points as shown on page 106 when conditions make the use of a short turnout with small radius necessary.

18" TRACK GAUGE

| | Frog * | | | Switch ** | | Actual Frog Lead | Closure Rails | | | |
| | | | | | | | Straight | Curved | | |
No.	Angle	Rail Weight (Lbs.)	Actual Toe Length	Length	Angle		Length	Length	M. O.	Radius
2¾	20 36'36"	12 to 35 40 to 48	1'-4½" 2'-0"	3'-6" 3'-6"	5°38'00" 5°38'00"	7'-7" 7'-4"	2'-8¼" 1'-9¾"	2'-10½" 2'-0"	1.13" 0.79"	11'-0" 7'-8"
3	18 55'29"	12 to 35 40 to 48	1'-4½" 2'-0"	5'-0" 5'-0"	3°56'30" 3°56'30"	9'-9" 9'-5"	3'-4¼" 2'-4¾"	3'-6¼" 2'-6¾"	1.38" 1.01"	13'-6" 9'-10"
3½	16 15'39"	12 to 35 40 to 48	1'-9" 2'-3"	5'-0" 5'-0"	3°56'30" 3°56'30"	10'-4" 10'-1"	3'-6¾" 2'-9¾"	3'-8½" 2'-11½"	1.20" 0.95"	17'-3" 13'-9"
4	14 15'00"	12 to 35 40 to 48	1'-9" 2'-3"	5'-0" 5'-0"	3°56'30" 3°56'30"	11'-2" 10'-11"	4'-4¾" 3'-7¾"	4'-6¼" 3'-9¼"	1.22" 1.02"	25'-2" 21'-0"
5	11 25'16"	12 to 35 40 to 48	1'-10½" 2'-6"	7'-6" 7'-6"	2°37'36" 2°37'36"	15'-7" 15'-4"	6'-2¼" 5'-3¾"	6'-3½" 5'-5"	1.45" 1.24"	41'-0" 35'-3"
6	9 31'38"	12 to 35 40 to 48	2'-3" 3'-0"	7'-6" 7'-6"	2°37'36" 2°37'36"	17'-0" 16'-8"	7'-2¾" 6'-1¾"	7'-3¾" 6'-2¾"	1.32" 1.12"	60'-9" 51'-9"
6	9 31'38"	12 to 35 40 to 48	2'-3" 3'-0"	10'-0" 10'-0"	1°58'12" 1°58'12"	19'-11" 19'-6"	7'-7¾" 6'-5¾"	7'-8¾" 6'-6¾"	1.55" 1.30"	58'-7" 49'-9"
7	8 10'17"	12 to 35 40 to 48	2'-6" 3'-4"	10'-0" 10'-0"	1°58'12" 1°58'12"	21'-5" 21'-0"	8'-10¾" 7'-7¾"	8'-11¾" 7'-8¾"	1.45" 1.25"	82'-11" 71'-5"
8	7 09'10"	12 to 35 40 to 48	2'-9" 3'-8"	10'-0" 10'-0"	1°58'12" 1°58'12"	22'-10" 22'-5"	10'-0¾" 8'-8¾"	10'-1½" 8'-9½"	1.37" 1.19"	1 11'-11" 97'-2"

*See page 108 for additional standard frog data.
**See page 110 for additional standard switch data.

"CARD" STANDARD TURNOUT DATA

STANDARD FROGS AND SWITCH POINTS WITH CLOSURE RAILS

24" TRACK GAUGE

Frog *				Switch **		Actual Frog Lead	Closure Rails			
							Straight	Curved		
No.	Angle	Rail Weight (Lbs.)	Actual Toe Length	Length	Angle		Length	Length	M. O.	Radius
2	28°04'21"	12 to 35	1'-5"	3'-6"	5°38'00"	8'-0"	3'-0¾"	3'-4¾"	1.99"	8'-8"
		40 to 48	1'-8"	3'-6"	5°38'00"	7'-10"	2'-7¾"	2'-11¾"	1.74"	7'-7"
2¼	25°03'25"	12 to 35	1'-4½"	3'-6"	5°38'00"	8'-7"	3'-8¼"	3'-11¾"	2.02"	11'-9"
		40 to 48	2'-0"	3'-6"	5°38'00"	8'-3"	2'-8¾"	3'-0¼"	1.53"	8'-11"
2½	22°37'12"	12 to 35	1'-4½"	3'-6"	5 38'00"	9'-2"	4'-3¼"	4'-6½"	2.02"	15'-4"
		40 to 48	2'-0"	3'-6"	5°38'00"	8'-10"	3'-3¾"	3'-7"	1.59"	12'-1"
2¾	20°36'36"	12 to 35	1'-4½"	3'-6"	5 38'00"	9'-9"	4'-10¼"	5'-1¼"	2.00"	19'-6"
		40 to 48	2'-0"	3'-6"	5°38'00"	9'-5"	3'-10¾"	4'-1¾"	1.62"	15'-10"
3	18°55'29"	12 to 35	1'-4½"	5'-0"	3°56'30"	12'-3"	5'-10¼"	6'-0¾"	2.37"	23'-2"
		40 to 48	2'-0"	5'-0"	3°56'30"	11'-10"	4'-9¾"	5'-0¼"	1.97"	19'-3"
3½	16°15'39"	12 to 35	1'-9"	5'-0"	3°56'30"	13'-2"	6'-4¾"	6'-7"	2.12"	30'-7"
		40 to 48	2'-3"	5'-0"	3°56'30"	12'-11"	5'-7¾"	5'-10"	1.88"	27'-1"
4	14°15'00"	12 to 35	1'-9"	5'-0"	3°56'30"	14'-3"	7'-5¾"	7'-7¾"	2.06"	42'-6"
		40 to 48	2'-3"	5'-0"	3°56'30"	14'-1"	6'-9¾"	6'-11¾"	1.88"	38'-10"
		50 to 60	2'-3"	5'-0"	4°25'12"	13'-7"	6'-3¾"	6'-5¾"	1.67"	37'-9"
5	11°25'16"	12 to 35	1'-10½"	7'-6"	2°37'36"	19'-8"	10'-3¼"	10'-4¾"	2.39"	67'-9"
		40 to 48	2'-6"	7'-6"	2°37'36"	19'-4"	9'-3¾"	9'-5¼"	2.17"	61'-6"
		50 to 60	2'-6"	7'-6"	2°56'48"	18'-10"	8'-9¾"	8'-11¼"	1.98"	60'-5"
6	9°31'38"	12 to 35	2'-3"	7'-6"	2°37'36"	21'-8"	11'-10¾"	12'-0¼"	2.17"	99'-10"
		40 to 48	3'-0"	7'-6"	2°37'36"	21'-4"	10'-9¾"	10'-11¼"	1.97"	90'-10"
		50 to 60	3'-0"	7'-6"	2°56'48"	20'-8"	10'-1¾"	10'-3¼"	1.77"	89'-5"
6	9°31'38"	12 to 35	2'-3"	10'-0"	1°58'12"	24'-10"	12'-6¾"	12'-8¼"	2.52"	96'-2"
		40 to 48	3'-0"	10'-0"	1 58'12"	24'-6"	11'-5¾"	11'-7¼"	2.30"	88'-0"
		50 to 60	3'-0"	10'-0"	2°12'30"	23'-10"	10'-9¾"	10'-11¼"	2.10"	85'-8"
7	8°10'17"	12 to 35	2'-6"	10'-0"	1°58'12"	27'-0"	14'-5¾"	14'-7"	2.36"	134'-9"
		40 to 48	3'-4"	10'-0"	1°58'12"	26'-7"	13'-2¾"	13'-4"	2.16"	123'-2"
		50 to 60	3'-4"	10'-0"	2°12'30"	25'-10"	12'-5¾"	12'-7"	1.97"	120'-11"
8	7°09'10"	12 to 35	2'-9"	10'-0"	1°58'12"	29'-1"	16'-3¾"	16'-4¾"	2.22"	181'-3"
		40 to 48	3'-8"	10'-0"	1°58'12"	28'-8"	14'-11¾"	15'-0¾"	2.04"	166'-6"
		50 to 60	3'-8"	10'-0"	2°12'30"	27'-9"	14'-0¾"	14'-1¾"	1.83"	163'-11"

*See page 108 for additional standard frog data.
**See page 110 for additional standard switch data.

"CARD" STANDARD TURNOUT DATA

STANDARD FROGS AND SWITCH POINTS WITH CLOSURE RAILS

30" TRACK GAUGE

| | Frog * | | | Switch ** | | Actual Frog Lead | Closure Rails | | | |
| | | | | | | | Straight | Curved | | |
No.	Angle	Rail Weight (Lbs.)	Actual Toe Length	Length	Angle		Length	Length	M.O.	Radius
1½	36°52'11"	12 to 35	1'-5"	3'-6"	5°38'00"	8'-0"	3'-0¾"	3'-7¼"	2.92"	6'-7"
1¾	31°53'25"	12 to 35	1'-5"	3'-6"	5°38'00"	8'-10"	3'-10¾"	4'-4¼"	3.00"	9'-6"
2	28°04'21"	12 to 35	1'-5"	3'-6"	5°38'00"	9'-8"	4'-8¾"	5'-1¾"	3.01"	13'-2"
		40 to 48	1'-8"	3'-6"	5°38'00"	9'-6"	4'-3¾"	4'-8¾"	2.77"	12'-1"
2¼	25°03'25"	12 to 35	1'-4½"	3'-6"	5°38'00"	10'-5"	5'-6¼"	5'-10¾"	2.99"	17'-5"
		40 to 48	2'-0"	3'-6"	5°38'00"	10'-1"	4'-6¾"	4'-11¼"	2.51"	14'-7"
2½	22°37'12"	12 to 35	1'-4½"	3'-6"	5°38'00"	11'-2"	6'-3¼"	6'-7¼"	2.94"	22'-4"
		40 to 48	2'-0"	3'-6"	5°38'00"	10'-10"	5'-3¾"	5'-7¾"	2.51"	19'-1"
2¾	20°36'36"	12 to 35	1'-4½"	5'-0"	3°56'30"	13'-11"	7'-6¼"	7'-9¾"	3.40"	26'-10"
		40 to 48	2'-0"	5'-0"	3°56'30"	13'-6"	6'-5¾"	6'-9¼"	2.95"	23'-3"
3	18°55'29"	12 to 35	1'-4½"	5'-0"	3°56'30"	14'-8"	8'-3¼"	8'-6½"	3.34"	32'-8"
		40 to 48	2'-0"	5'-0"	3°56'30"	14'-4"	7'-3¾"	7'-7"	2.97"	29'-0"
		50 to 60	2'-0"	5'-0"	4°25'12"	14'-0"	6'-11¾"	7'-3"	2.75"	28'-8"
3½	16°15'39"	12 to 35	1'-9"	5'-0"	3°56'30"	16'-0"	9'-2¾"	9'-5½"	3.05"	44'-0"
		40 to 48	2'-3"	5'-0"	3°56'30"	15'-9"	8'-5¾"	8'-8½"	2.81"	40'-6"
		50 to 60	2'-3"	5'-0"	4°25'12"	15'-3"	7'-11¾"	8'-2½"	2.54"	39'-8"
4	14°15'00"	12 to 35	1'-9"	5'-0"	3°56'30"	17'-5"	10'-7¾"	10'-10¼"	2.92"	60'-4"
		40 to 48	2'-3"	5'-0"	3°56'30"	17'-2"	9'-10¾"	10'-1¼"	2.72"	56'-2"
		50 to 60	2'-3"	5'-0"	4°25'12"	16'-8"	9'-4¾"	9'-7¼"	2.47"	56'-0"
5	11°25'16"	12 to 35	1'-10½"	7'-6"	2°37'36"	23'-9"	14'-4¼"	14'-6¼"	3.34"	94'-7"
		40 to 48	2'-6"	7'-6"	2°37'36"	23'-5"	13'-4¾"	13'-6¾"	3.12"	88'-4"
		50 to 60	2'-6"	7'-6"	2°56'48"	22'-9"	12'-8¾"	12'-10¾"	2.86"	87'-2"
6	9°31'38"	12 to 35	2'-3"	7'-6"	2°37'36"	26'-4"	16'-6¾"	16'-8½"	3.01"	138'-9"
		40 to 48	3'-0"	7'-6"	2°37'36"	26'-0"	15'-5¾"	15'-7½"	2.82"	129'-9"
		50 to 60	3'-0"	7'-6"	2°56'48"	25'-3"	14'-8¾"	14'-10½"	2.56"	129'-6"
6	9°31'38"	12 to 35	2'-3"	10'-0"	1°58'12"	29'-10"	17'-6¾"	17'-8½"	3.51"	134'-3"
		40 to 48	3'-0"	10'-0"	1°58'12"	29'-5"	16'-4¾"	16'-6½"	3.28"	125'-5"
		50 to 60	3'-0"	10'-0"	2°12'30"	28'-8"	15'-7¾"	15'-9½"	3.03"	123'-8"
7	8°10'17"	12 to 35	2'-6"	10'-0"	1°58'12"	32'-8"	20'-1¾"	20'-3¼"	3.28"	187'-3"
		40 to 48	3'-4"	10'-0"	1°58'12"	32'-3"	18'-10¾"	19'-0¼"	3.08"	175'-9"
		50 to 60	3'-4"	10'-0"	2°12'30"	31'-4"	17'-11¾"	18'-1¼"	2.84"	173'-11"
8	7°09'10"	12 to 35	2'-9"	10'-0"	1°58'12"	35'-4"	22'-6¾"	22'-8"	3.07"	250'-7"
		40 to 48	3'-8"	10'-0"	1°58'12"	34'-11"	21'-2¾"	21'-4"	2.89"	235'-10"
		50 to 60	3'-8"	10'-0"	2°12'30"	33'-11"	20'-2¾"	20'-4"	2.63"	235'-7"

*See page 108 for additional standard frog data.
**See page 110 for additional standard switch data.

"CARD" STANDARD TURNOUT DATA

STANDARD FROGS AND SWITCH POINTS WITH CLOSURE RAILS

36" TRACK GAUGE

No.	Frog Angle	Frog Rail Weight (Lbs.)	Frog Actual Toe Length	Switch Length	Switch Angle	Actual Frog Lead	Closure Rails Straight Length	Closure Rails Curved Length	Closure Rails Curved M.O.	Closure Rails Curved Radius
1¼	43°36′11″	12 to 35	1′–5″	3′–6″	5°38′00″	8′–3″	3′–3¾″	4′–1″	4.02″	6′–2″
1½	36°52′11″	12 to 35	1′–5″	3′–6″	5°38′00″	9′–3″	4′–3¾″	4′–11½″	4.02″	9′–1″
1¾	31°53′25″	12 to 35	1′–5″	3′–6″	5°38′00″	10′–4″	5′–4¾″	5′–11½″	4.07″	13′–0″
2	28°04′21″	12 to 35	1′–5″	3′–6″	5°38′00″	11′–4″	6′–4¾″	6′–10¾″	4.03″	17′–7″
		40 to 48	1′–8″	3′–6″	5°38′00″	11′–2″	5′–11¾″	6′–5¾″	3.80″	16′–7″
2¼	25°03′25″	12 to 35	1′–4½″	3′–6″	5°38′00″	12′–3″	7′–4¼″	7′–9½″	3.96″	23′–0″
		40 to 48	2′–0″	3′–6″	5°38′00″	11′–11″	6′–4¾″	6′–10″	3.47″	20′–2″
2½	22°37′12″	12 to 35	1′–4½″	5′–0″	3°56′30″	15′–3″	8′–10¼″	9′–3″	4.52″	28′–5″
		40 to 48	2′–0″	5′–0″	3°56′30″	14′–10″	7′–9¾″	8′–2½″	4.00″	25′–2″
		50 to 60	2′–0″	5′–0″	4°25′12″	14′–6″	7′–5¾″	7′–10½″	3.75″	24′–10″
2¾	20°36′36″	12 to 35	1′–4½″	5′–0″	3°56′30″	16′–3″	9′–10¼″	10′–2½″	4.45″	35′–1″
		40 to 48	2′–0″	5′–0″	3°56′30″	15′–10″	8′–9¾″	9′–2″	3.99″	31′–6″
3	18°55′29″	12 to 35	1′–4½″	5′–0″	3°56′30″	17′–2″	10′–9¼″	11′–1¼″	4.35″	42′–6″
		40 to 48	2′–0″	5′–0″	3°56′30″	16′–10″	9′–9¾″	10′–1¾″	3.97″	38′–10″
		50 to 60	2′–0″	5′–0″	4°25′12″	16′–5″	9′–4¾″	9′–8¾″	3.69″	38′–5″
3½	16°15′39″	12 to 35	1′–9″	5′–0″	3°56′30″	18′–9″	11′–11¾″	12′–3¼″	3.96″	57′–1″
		40 to 48	2′–3″	5′–0″	3°56′30″	18′–6″	11′–2¾″	11′–6¼″	3.72″	53′–7″
		50 to 60	2′–3″	5′–0″	4°25′12″	18′–0″	10′–8¾″	11′–0¼″	3.41″	53′–3″
4	14°15′00″	12 to 35	1′–9″	5′–0″	3°56′30″	20′–6″	13′–8¾″	13′–11¾″	3.77″	77′–8″
		40 to 48	2′–3″	5′–0″	3°56′30″	20′–4″	13′–0¾″	13′–3¾″	3.59″	74′–0″
		50 to 60	2′–3″	5′–0″	4°25′12″	19′–9″	12′–5¾″	12′–8¾″	3.28″	74′–2″
5	11°25′16″	12 to 35	1′–10½″	7′–6″	2°37′36″	27′–9″	18′–4¼″	18′–6¾″	4.26″	120′–11″
		40 to 48	2′–6″	7′–6″	2°37′36″	27′–6″	17′–5¾″	17′–8¼″	4.07″	115′–3″
		50 to 60	2′–6″	7′–6″	2°56′48″	26′–9″	16′–8¾″	16′–11¼″	3.75″	114′–6″
6	9°31′38″	12 to 35	2′–3″	7′–6″	2°37′36″	31′–1″	21′–3¾″	21′–5¾″	3.87″	178′–4″
		40 to 48	3′–0″	7′–6″	2°37′36″	30′–9″	20′–2¾″	20′–4¾″	3.68″	169′–4″
		50 to 60	3′–0″	7′–6″	2°56′48″	29′–10″	19′–3¾″	19′–5¾″	3.36″	169′–7″
6	9°31′38″	12 to 35	2′–3″	10′–0″	1°58′12″	34′–9″	22′–5¾″	22′–7¾″	4.49″	171′–8″
		40 to 48	3′–0″	10′–0″	1°58′12″	34′–5″	21′–4¾″	21′–6¾″	4.28″	163′–6″
		50 to 60	3′–0″	10′–0″	2°12′30″	33′–7″	20′–6¾″	20′–8¾″	3.97″	162′–3″
7	8°10′17″	12 to 35	2′–6″	10′–0″	1°58′12″	38′–3″	25′–8¾″	25′–10½″	4.19″	239′–1″
		40 to 48	3′–4″	10′–0″	1°58′12″	37′–11″	24′–6¾″	24′–8½″	4.00″	228′–3″
		50 to 60	3′–4″	10′–0″	2°12′30″	36′–10″	23′–5¾″	23′–7½″	3.70″	227′–0″
8	7°09′10″	12 to 35	2′–9″	10′–0″	1°58′12″	41′–7″	28′–9¾″	28′–11¼″	3.91″	319′–10″
		40 to 48	3′–8″	10′–0″	1°58′12″	41′–2″	27′–5¾″	27′–7¼″	3.73″	305′–2″
		50 to 60	3′–8″	10′–0″	2°12′30″	40′–0″	26′–3¾″	26′–5¼″	3.42″	306′–4″

*See page 108 for additional standard frog data.
**See page 110 for additional standard switch data.

"CARD" STANDARD TURNOUT DATA

STANDARD FROGS AND SWITCH POINTS WITH CLOSURE RAILS

42″ TRACK GAUGE

Frog *				Switch **		Actual Frog Lead	Closure Rails			
No.	Angle	Rail Weight (Lbs.)	Actual Toe Length	Length	Angle		Straight Length	Curved Length	Curved M.O.	Curved Radius
1¼	43°36′11″	12 to 35	1′-5″	3′-6″	5°38′00″	9′-4″	4′-4¾″	5′-3¾″	5.22″	8′-0″
1½	36°52′11″	12 to 35	1′-5″	3′-6″	5°38′00″	10′-7″	5′-7¾″	6′-4¾″	5.21″	11′-9″
1¾	31°53′25″	12 to 35	1′-5″	3′-6″	5°38′00″	11′-9″	6′-9¾″	7′-5¾″	5.12″	16′-4″
2	28°04′21″	12 to 35	1′-5″	3′-6″	5°38′00″	12′-11″	7′-11¾″	8′-6¾″	5.00″	21′-10″
		40 to 48	1′-8″	3′-6″	5°38′00″	12′-10″	7′-7¾″	8′-2¾″	4.81″	21′-0″
2¼	25°03′25″	12 to 35	1′-4½″	3′-6″	5°38′00″	14′-1″	9′-2¼″	9′-8½″	4.93″	28′-8″
		40 to 48	2′-0″	3′-6″	5°38′00″	13′-9″	8′-2¾″	8′-9″	4.44″	25′-10″
2½	22°37′12″	12 to 35	1′-4½″	5′-0″	3°56′30″	17′-4″	10′-11¼″	11′-4¾″	5.56″	35′-0″
		40 to 48	2′-9″	5′-0″	3°56′30″	17′-0″	9′-11¾″	10′-5¼″	5.09″	32′-0″
		50 to 60	2′-0″	5′-0″	4°25′12″	16′-7″	9′-6¾″	10′-0¼″	4.77″	31′-7″
2¾	20°36′36″	12 to 35	1′-4½″	5′-0″	3°56′30″	18′-6″	12′-1¼″	12′-6¼″	5.45″	43′-0″
		40 to 48	2′-0″	5′-0″	3°56′30″	18′-2″	11′-1¾″	11′-6¾″	5.04″	39′-9″
3	18°55′29″	12 to 35	1′-4½″	5′-0″	3°56′30″	19′-8″	13′-3¼″	13′-7¾″	5.35″	52′-3″
		40 to 48	2′-0″	5′-0″	3°56″30″	19′-4″	12′-3¾″	12′-8¼″	4.97″	48′-7″
		50 to 60	2′-0″	5′-0″	4°25′12″	18′-10″	11′-9¾″	12′-2¼″	4.62″	48′-2″
3½	16°15′39″	12 to 35	1′-9″	5′-0″	3°56′30″	21′-7″	14′-9¾″	15′-1¾″	4.88″	70′-5″
		40 to 48	2′-3″	5′-0″	3°56′30″	21′-4″	14′-0¾″	14′-4¾″	4.64″	66′-11″
		50 to 60	2′-3″	5′-0″	4°25′12″	20′-9″	13′-5¾″	13′-9¾″	4.28″	66′-9″
4	14°15′00″	12 to 35	1′-9″	5′-0″	3°56′30″	23′-8″	16′-10¾″	17′-2¼″	4.63″	95′-6″
		40 to 48	2′-3″	5′-0″	3°56′30″	23′-5″	16′-1¾″	16′-5¼″	4.43″	91′-4″
		50 to 60	2′-3″	5′-0″	4°25′12″	22′-9″	15′-5¾″	15′-9¼″	4.06″	91′-11″
5	11°25′16″	12 to 35	1′-10½″	7′-6″	2°37′36″	31′-10″	22′-5¼″	22′-8″	5.21″	147′-8″
		40 to 48	2′-6″	7′-6″	2°37′36″	31′-6″	21′-5¾″	21′-8½″	4.99″	141′-5″
		50 to 60	2′-6″	7′-6″	2°56′48″	30′-9″	20′-8¾″	20′-11½″	4.64″	141′-8″
6	9°31′38″	12 to 35	2′-3″	7′-6″	2°37′36″	35′-9″	25′-11¾″	26′-2¼″	4.72″	217′-5″
		40 to 48	3′-0″	7′-6″	2°37′36″	35′-5″	24′-10¾″	25′-1¼″	4.53″	208′-5″
		50 to 60	3′-0″	7′-6″	2°56′48″	34′-5″	23′-10¾″	24′-1¼″	4.16″	209′-11″
6	9°31′38″	12 to 35	2′-3″	10′-0″	1°58′12″	39′-9″	27′-5¾″	27′-8¼″	5.49″	209′-11″
		40 to 48	3′-0″	10′-0″	1°58′12″	39′-4″	26′-3¾″	26′-6¼″	5.26″	201′-1″
		50 to 60	3′-0″	10′-0″	2°12′30″	38′-5″	25′-4¾″	25′-7¼″	4.91″	200′-5″
7	8°10′17″	12 to 35	2′-6″	10′-0″	1°58′12″	43′-11″	31′-4¾″	31′-6¾″	5.11″	291′-7″
		40 to 48	3′-4″	10′-0″	1°58′12″	43′-6″	30′-1¾″	30′-3¾″	4.91″	280′-1″
		50 to 60	3′-4″	10′-0″	2°12′30″	42′-4″	28′-11¾″	29′-1¾″	4.57″	280′-0″
8	7°09′10″	12 to 35	2′-9″	10′-0″	1°58′12″	47′-10″	35′-0¾″	35′-2½″	4.76″	389′-2″
		40 to 48	3′-8″	10′-0″	1°58′12″	47′-6″	33′-9¾″	33′-11½″	4.59″	375′-4″
		50 to 60	3′-8″	10′-0″	2°12′30″	46′-1″	32′-4¾″	32′-6½″	4.21″	377′-0″

*See page 108 for additional standard frog data.
**See page 110 for additional standard switch data.

"CARD" STANDARD RAIL TURNOUTS

CURVED FROG AND SWITCH POINTS

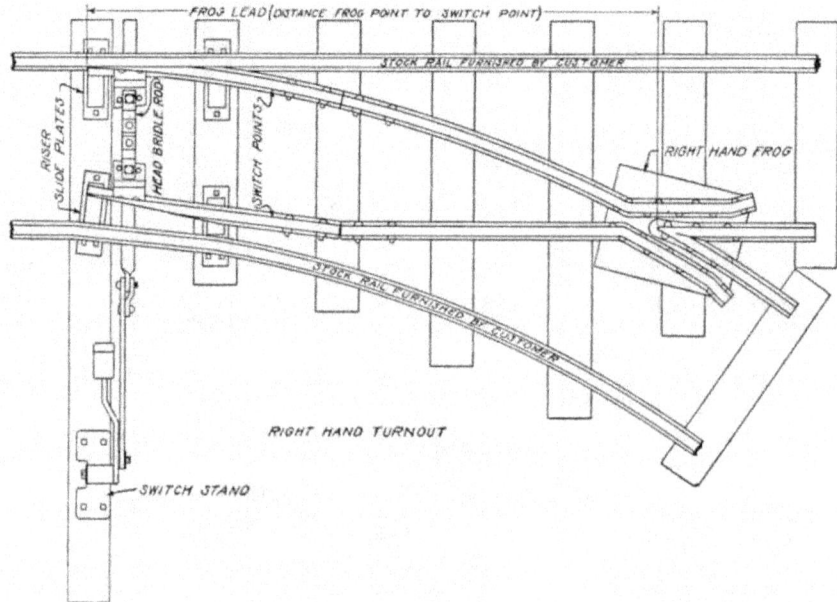

When, for lack of room, short leads are necessary, we recommend this style of turnout.

One wing rail of the frog and one switch point are curved to the radius of the turnout. Note that the frog wing rails are made long enough to abut the switch points, doing away with short closure rails and saving two pairs of splice bars. The cutting, drilling and bending of the closure rails is also saved. Considerable saving is made in laying the turnout as the lead and radius are established.

This style of turnout can be furnished of any weight rail, to any radius or track gauge. However, where the lead becomes too great, the frog becomes too long to conveniently handle, and we recommend the use of a turnout with straight frog and switch points, using closure rails.

This style turnout is not reversible and must be specified as either right or left hand. A right hand turnout is illustrated.

Data and price lists for these turnouts on page 107.

"CARD" STANDARD RAIL TURNOUTS
CURVED FROG AND SWITCH POINTS

A Complete Turnout consists of:

> One Curved Frog.
> One Pair Switch Points with Clips.
> Bridle Rod.
> Necessary Riser Plates.
> Plain Ground Throw Switch Stand No. 1620.

Switch Stand may be omitted, or other types furnished if desired.

Refer to page 106 for illustration of this turnout.

As this turnout is not reversible, it is necessary that you specify the quantity of right hand and of left hand turnouts required.

Prices quoted on sizes not listed, if given weight of rail, track gauge and frog number; or the lead or centerline radius required.

In ordering from this table, please specify the weight of rail, track gauge and frog number.

Specifications; Prices and Average Weights in Pounds, Each Complete Turnout

Track Gauge	Frog Number	Switch Point Length	Actual Frog Lead	PRICE Rail Size					WEIGHT Rail Size				
				12 Lb.	16 Lb.	20 Lb.	25 Lb.	30 Lb.	12 Lb.	16 Lb.	20 Lb.	25 Lb.	30 Lb.
18"	1½	3'-0"	4'-7"	$26.85	$29.51	$32.31	$39.10	$43.42	120	153	173	219	258
	1¾	3'-0"	5'-1"	27.04	29.73	32.58	39.40	43.77	124	158	179	227	267
	2	3'-0"	5'-7"	27.20	29.94	32.83	39.71	44.15	127	163	186	234	276
	2¼	3'-0"	6'-1"	28.37	31.14	34.22	41.16	45.71	137	176	201	254	300
	2½	3'-0"	6'-6"	28.52	31.77	34.88	41.96	46.55	141	181	207	261	308
	2¾	3'-6"	7'-0"	28.80	32.83	35.82	42.97	47.68	146	188	215	271	320
	3	3'-6"	7'-5"	28.94	33.45	36.57	43.76	48.43	150	192	220	278	328
	3	5'-0"	8'-6"	30.63	35.28	38.66	46.01	51.04	161	208	238	299	355
	3½	5'-0"	9'-5"	31.98	36.89	40.90	48.93	54.76	181	233	269	336	400
	4	5'-0"	10'-3"	32.46	37.41	42.69	51.70	57.36	187	242	280	349	416
24"	1¼	3'-0"	5'-2"	27.61	29.46	33.22	40.09	44.51	127	162	184	232	273
	1½	3'-0"	5'-11"	27.86	30.67	33.59	40.55	45.06	132	170	193	244	287
	1¾	3'-0"	6'-7"	28.09	30.95	33.94	40.95	45.53	137	176	201	254	299
	2	3'-6"	7'-3"	28.33	31.45	34.31	41.35	46.02	142	183	209	264	311
	2¼	3'-6"	7'-11"	29.55	32.68	35.78	42.90	47.71	153	198	228	286	338
	2½	3'-6"	8'-7"	29.78	33.41	36.58	43.85	48.75	158	205	236	297	351
	2½	5'-0"	9'-7"	31.43	35.20	38.64	46.03	51.30	169	220	254	317	376
	2¾	5'-0"	10'-3"	31.77	36.17	39.64	47.18	52.55	176	229	264	330	391
	3	5'-0"	11'-0"	32.05	36.95	40.56	48.15	53.53	182	236	273	342	406

"CARD" STANDARD FROGS

We illustrate the two styles of frogs that we regularly furnish, made from 12 lb. to 60 lb. rail. Any weight rail can be furnished.

Our standard frogs follow the standards approved by the American Mining Congress and American Standards Association.

The heavy steel plate, on which they are mounted, extends well ahead of the frog point, giving additional support and strength to the wing rails. Lighter weight rails are welded and heavier weight rails are riveted to the plate.

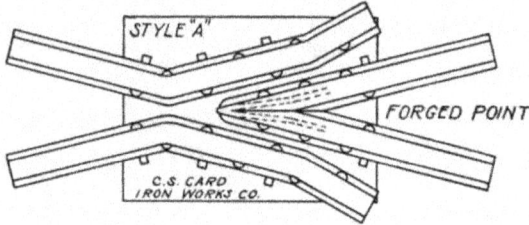

Style "A" Frogs have forged points and are used with the lighter weight rails and in the lower frog numbers. The point rails are welded after forging, making practically a solid frog point.

Style "B" Frogs have machined points and are used with the heavier weight rails and in the higher frog numbers. The short and long point rails are welded together, making practically a solid one-piece frog point.

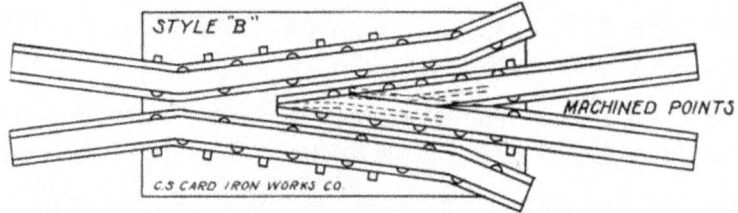

See page 109 for price list of these frogs.

Our standard frog specifications are given below. See illustration on page 97 for names of frog parts and additional frog data.

Standard Frog Specifications

Frog Number	Frog Angle	Spread in 12"	Standard Frog Lengths					
			12 Lb. to 35 Lb. Rail ¼" Frog Point			40 Lb. to 60 Lb. Rail ⅜" Frog Point		
			Length		Total Length	Length		Total Length
			Toe	Heel		Toe	Heel	
1¼	43°36′11″	9.60″	1′–5″	1′–10″	3′–3″	....	....	....
1½	36°52′11″	8.00″	1′–5″	1′–10″	3′–3″	....	....	....
1¾	31°53′25″	6.86″	1′–5″	1′–10″	3′–3″	....	....	....
2	**28°04′21″**	**6.00″**	**1′–5″**	**1′–10″**	**3′–3″**	**1′–8″**	**2′–4″**	**4′–0″**
2¼	25°03′25″	5.33″	1′–4½″	2′–1½″	3′–6″	2′–0″	3′–0″	5′–0″
2½	22°37′12″	4.80″	1′–4½″	2′–1½″	3′–6″	2′–0″	3′–0″	5′–0″
2¾	20°36′36″	4.36″	1′–4½″	2′–1½″	3′–6″	2′–0″	3′–0″	5′–0″
3	**18°55′29″**	**4.00″**	**1′–4½″**	**2′–1½″**	**3′–6″**	**2′–0″**	**3′–0″**	**5′–0″**
3½	16°15′39″	3.43″	1′–9″	2′–9″	4′–6″	2′–3″	3′–9″	6′–0″
4	**14°15′00″**	**3.00″**	**1′–9″**	**2′–9″**	**4′–6″**	**2′–3″**	**3′–9″**	**6′–0″**
4½	12°40′50″	2.67″	1′–10½″	3′–1½″	5′–0″	2′–6″	4′–6″	7′–0″
5	**11°25′16″**	**2.40″**	**1′–10½″**	**3′–1½″**	**5′–0″**	**2′–6″**	**4′–6″**	**7′–0″**
6	**9°31′38″**	**2.00″**	**2′–3″**	**3′–9″**	**6′–0″**	**3′–0″**	**5′–0″**	**8′–0″**
7	**8°10′17″**	**1.71″**	**2′–6″**	**4′–3″**	**6′–9″**	**3′–4″**	**5′–8″**	**9′–0″**
8	**7°09′10″**	**1.50″**	**2′–9″**	**4′–9″**	**7′–6″**	**3′–8″**	**6′–4″**	**10′–0″**

Frogs in bold face type are American Mining Congress Standard sizes.
All standard 16 lb. to 60 lb. frogs have 1⅝" wide flangeways; 12 lb. frogs have 1½" wide flangeways.

"CARD" STANDARD FROGS

The following table covers our standard frogs as shown on page 108.

Those listed in bold face type have been approved by the American Mining Congress and American Standards Association as American Standards and have been recommended for use as follows, based on 42″ gauge:—

No. 2, No. 2½ and No. 3 Frogs for room turnouts.

No. 3, No. 4, No. 5 and No. 6 frogs for turnouts for main haulage.

No. 5 and No. 6 Frogs for main line haulage where traffic is fast.

We can furnish you frogs of any frog number or weight of rail required, though not specified in this list.

See page 108 for dimensions and data covering frogs listed below.

Frogs marked with an asterisk (*) are carried in stock for immediate shipment.

Specifications; Prices and Average Weights in Pounds, Each

FROG NUMBER		RAIL SIZE							
		12 Lb.	16 Lb.	20 Lb.	25 Lb.	30 Lb.	40 Lb.	45 Lb.	60 Lb.
1¼	Price	$ 9.26	$10.35	$11.70	$14.12	$15.98			
	Weight	52	69	81	98	118			
1½	Price	$ 9.26	$10.35	$11.70	$14.12	15.96			
	Weight	52	69	81	98	$118			
1¾	Price	$ 9.27	$10.36	$11.70	$14.12	$15.96			
	Weight	52	69	81	97	118			
2	Price	$ 9.27	$10.36*	$11.70*	$14.12	$15.96	$21.81		
	Weight	52	69	81	97	118	191		
2¼	Price	$ 9.98	$11.15	$12.53	$15.04	$17.02	$23.79		
	Weight	58	76	90	108	131	231		
2½	Price	$ 9.98	$11.61	$12.97*	$15.47	$17.47*	$24.25	$27.31	$36.63
	Weight	58	76	90	108	131	230	260	331
2¾	Price	$10.11	$12.18	$13.56	$16.09	$18.11	$24.74	$27.79	$37.10
	Weight	60	79	92	111	135	230	260	331
3	Price	$10.11	$12.64*	$14.01*	$16.54	$18.56*	$25.27*	$28.33	$37.62
	Weight	60	79	92	111	135	231	261	333
3½	Price	$11.30	$14.01	$15.96	$18.67	$20.98	$27.91	$31.23	$41.95
	Weight	75	98	115	139	169	272	308	392
4	Price	$11.28	$13.99	$15.94	$19.14	$21.85*	$30.34*	$33.75	$44.62*
	Weight	75	98	115	140	169	286	323	410
5	Price	$12.38	$15.12	$18.03	$21.74	$26.81	$35.33	$38.90	$50.22
	Weight	79	103	122	147	179	319	360	458
6	Price	$14.07	$17.03	$21.87	$25.22	$31.03*	$39.92*	$43.75	$56.47*
	Weight	95	125	146	177	214	363	410	520
7	Price					$39.47	$48.91	$53.87	$64.38
	Weight					243	404	457	578
8	Price					$47.81	$57.98	$63.21	$74.19
	Weight					269	448	507	639

For 35-lb. Rail Frogs, Add 15% to 30-lb. Prices.
Style "A" Frogs are standard above heavy line.
Style "B" Frogs are standard below heavy line.

"CARD" STANDARD SPLIT SWITCHES

On orders for complete split switches, we regularly furnish all parts as shown and listed below. They are designed for heavy duty.

All points are accurately machined to rail template. The switch point bridle rod clips have a series of holes on an angle, for attaching the head and back rods, permitting some adjustment.

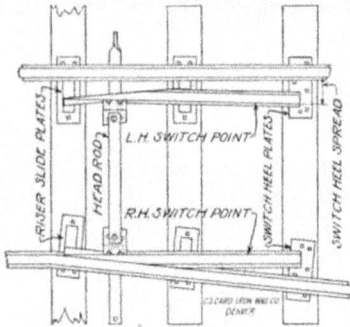

See page 111 for prices of these switches. In ordering specify the switch point length, weight of rail and track gauge.

**SPLIT SWITCHES, 3'-0"; 3'-6"
AND 4'-0" LONG**

Complete Split Switch consists of:—
One Pair of Switch Points with Clips;
One Head Rod; Two Pairs of Riser and
One Pair of Heel Plates.

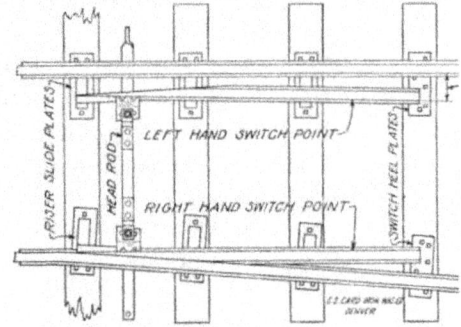

SPLIT SWITCHES, 5'-0" AND 6'-0" LONG.

Complete Split Switch consists of:—
One Pair of Switch Points with Clips;
One Head Rod; Three Pairs of Riser and
One Pair of Heel Plates.

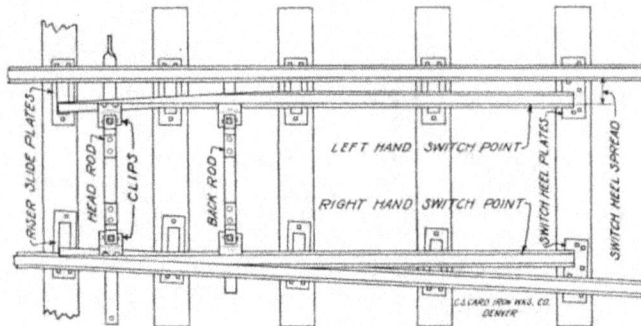

SPLIT SWITCH, 7'-6" LONG.

Complete Split Switch Consists of:—
One Pair of Switch Points with Clips;
One Head and One Back Rod; Four Pairs
of Riser and One Pair of Heel Plates.

**SPLIT SWITCH, 10'-0"
LONG.**

Complete Split Switch consists of:—

One Pair of Switch Points with Clips; One Head and Two Back Rods; Five Pairs of Riser and One Pair of Heel Plates.

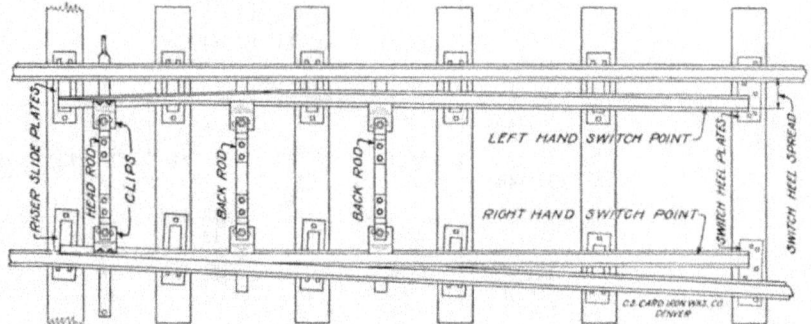

Head and back rods will be furnished for American Mining Congress standard $3\frac{1}{2}$" throw, unless otherwise specified.

American Mining Congress standard lengths are 3'-6", 5'-0", 7'-6" and 10'-0" long, and we recommend their use wherever possible.

For the above split switches, "CARD" Standard "Switch Heel Spread" for 3'-0" switches, 12 lb. to 48 lb. rail, is 4"; for 3'-6" and longer switches, 12 lb. to 48 lb. rail, it is $4\frac{1}{2}$"; for 50 lb. to 60 lb. rail, it is 5".

"CARD" STANDARD SPLIT SWITCHES

In the table below are listed complete split switches as shown on page 110.

American Mining Congress standard lengths are shown in bold face type.

Switches marked with an asterisk (*) are carried in stock for immediate shipment.

In ordering, please specify the switch point length, weight of rail and track gauge.

See page 112 for the parts which go to make up these complete split switches. Any of the parts may be ordered separately.

Specifications; Prices and Average Weights in Pounds, Each Complete Split Switch.

SWITCH LENGTH		RAIL SIZE							
		12 Lb.	16 Lb.	20 Lb.	25 Lb.	30 Lb.	40 Lb.	45 Lb.	60 Lb.
3'–0"	Price	$ 8.40	$ 9.52	$10.13	$10.91	$12.77	$14.65	. . . .	. . . .
	Weight	49	68	75	85	105	127	. . . .	. . . .
3'–6"	Price	$ 8.57	$ 9.74*	$10.40*	$11.22	$13.16*	$15.16	$16.48	$23.32
	Weight	53	73	81	92	114	139	150	197
4'–0"	Price	$ 9.10	$10.31	$11.02	$11.91	$13.90	$16.04	$17.52	$24.44
	Weight	57	78	87	100	123	151	163	217
5'–0"	Price	$10.30	$11.75*	$12.56*	$13.55	$15.81*	$18.32*	$20.50	$27.69*
	Weight	68	94	105	121	149	185	200	271
6'–0"	Price	$12.26	$13.79	$14.69	$15.78	$17.81	$20.23	$22.54	$30.12
	Weight	75	103	117	136	167	209	227	307
7'–6"	Price	$16.56	$18.49	$19.63	$20.98	$24.23*	$27.48*	$29.80	$38.78*
	Weight	105	143	161	184	226	279	302	408
10'–0"	Price	. . . .	. . . .	$27.04	$28.68	$32.28	$36.48	$40.00	$50.78
	Weight	. . . .	. . . .	216	247	303	373	404	544
6" Gauge differential for 3'–0"to 6'–0" switch	Price	$ 0.08	$ 0.09	$ 0.09	$ 0.09	$ 0.10	$ 0.10	$ 0.10	$ 0.14
	Weight	1	2	2	2	2	2	2	3
6" Gauge differential for 7'–6" switch	Price	$ 0.16	$ 0.18	$ 0.18	$0 .18	$ 0.20	$ 0.20	$ 0.20	$ 0.28
	Weight	3	3	3	3	4	4	4	5
6" Gauge differential for 10'–0" switch	Price	. . . .	. . . .	$ 0.27	$ 0.27	$ 0.30	$ 0.30	$ 0.30	$ 0.42
	Weight	. . . .	. . . .	5	5	6	6	6	8

Prices and weights given are for complete split switches for 36" gauge. For other than 36" gauge, add to or subtract from these figures, the differential given for each 6" change in gauge.

For 35 lb. split switches, add 15% to 30 lb. prices.

"CARD" STANDARD SPLIT SWITCH PARTS

The tables below list all the parts that are used in the standard split switches as shown on pages 110 and 111.

Split Switch Points with Bridle Rod Clips

Specifications; Prices and Average Weights in Pounds, Each Pair.

SWITCH LENGTH		RAIL SIZE							
		12 Lb.	16 Lb.	20 Lb.	25 Lb.	30 Lb.	40 Lb.	45 Lb.	60 Lb.
3'-0"	Price	$ 5.25	$ 5.75	$ 6.37	$ 7.07	$ 8.26	$ 9.78	....	....
	Weight	24	32	39	48	58	77		
3'-6"	Price	$ 5.43	$ 5.95*	$ 6.64*	$ 7.38	$ 8.65*	$10.30	$11.62	$17.26
	Weight	28	37	45	56	67	89	100	128
4'-0"	Price	$ 5.95	$ 6.54	$ 7.27	$ 8.06	$ 9.39	$11.17	$12.65	$18.40
	Weight	32	42	51	63	76	101	113	148
5'-0"	Price	$ 6.67	$ 7.33*	$ 8.16*	$ 9.06	$10.52*	$12.53*	$14.73	$20.49*
	Weight	39	51	63	78	94	125	140	188
6'-0"	Price	$ 8.65	$ 9.37	$10.30	$11.31	$12.52	$14.46	$16.77	$22.92
	Weight	46	61	75	93	112	149	167	224
7'-6"	Price	$11.05	$11.77	$12.92	$14.20	$16.24*	$18.85*	$21.15	$28.13*
	Weight	60	79	97	119	143	191	213	287
10'-0"	Price	....	....	$18.02	$19.58	$21.59	$24.96	$28.48	$36.67
	Weight			130	160	193	256	286	385

For 35-lb. Switch Points, Add 15% to 30-lb. Prices.
American Mining Congress Standard Lengths in Bold Face Type.
*Carried in stock for immediate shipment.

Bridle Rods and Switch Plates

Specifications; Prices and Average Weights in Pounds.

		RAIL SIZE				
		12 Lb.	16 to 25 Lb.	30 to 35 Lb.	40 to 48 Lb.	50 to 60 Lb.
Head Rod, 36" Gauge		$1\frac{1}{2}$" x $\frac{1}{2}$"	2" x $\frac{1}{2}$"	2" x $\frac{5}{8}$"	2" x $\frac{5}{8}$"	$2\frac{1}{2}$" x $\frac{5}{8}$"
	Price, each	$1.66	$1.92	$2.20	$2.25	$2.60
	Weight, each	13	17	21	21	27
Back Rod, 36" Gauge	Price, each	$1.42	$1.66	$1.92	$1.96	$2.29
	Weight, each	12	16	19	19	25
6" Gauge Differential	Price, each	$0.08	$0.09	$0.10	$0.10	$0.14
(Head and Back Rods)	Weight, each	1.3	1.7	2.1	2.1	2.7
Switch Side Plates, $\frac{3}{16}$" Riser		$3\frac{1}{2}$" x $\frac{3}{16}$"	4" x $\frac{1}{4}$"	4" x $\frac{5}{16}$"	....	....
	Price, per pair	$0.48	$0.64	$0.78		
	Weight, per pair	4	$6\frac{1}{2}$	$8\frac{1}{2}$		
Switch Slide Plates, $\frac{1}{4}$" Riser		....	....	....	$4\frac{1}{2}$" x $\frac{5}{16}$"	5" x $\frac{3}{8}$"
	Price, per pair				$0.91	$1.16
	Weight, per pair				10	14
Switch Heel Plates		4" x $\frac{3}{16}$"	$4\frac{1}{2}$" x $\frac{1}{4}$"	$4\frac{1}{2}$" x $\frac{3}{16}$"	$4\frac{1}{2}$" x $\frac{5}{16}$"	5" x $\frac{3}{8}$"
	Price, per pair	$0.49	$0.59	$0.75	$0.81	$1.14
	Weight, per pair	4	6	8	9	14
Bridle Rod Clips		$3\frac{1}{2}$" x $\frac{1}{4}$"	4" x $\frac{1}{4}$"	4" x $\frac{5}{16}$"	4" x $\frac{3}{8}$"	4" x $\frac{7}{16}$"
	Price, per pair	$0.56	$0.60	$0.69	$0.74	$0.88
	Weight, per pair	2	3	$3\frac{1}{2}$	$4\frac{1}{2}$	7

All parts listed in this table are carried in stock for immediate shipment.

"CARD" GROUND THROW SWITCH STANDS

STAND WITH PLAIN CONNECTING ROD.

A well made, substantial ground throw that is popular with the mining and industrial trade. All parts are steel forgings except the base plate and lever weight. All wearing parts are machine finished.

Where the switch is subjected to heavy and frequent service we recommend the throw with the spring connecting rod shown below. The rod and spring can be adjusted so that the switch points will be held tight with equal pressure against either rail; resulting in fewer broken points and derailments caused by splitting the switch points.

Unless otherwise specified, we will furnish these ground throws with the standard throw of $3\frac{1}{2}''$.

All ground throws listed are carried in stock for immediate shipment.

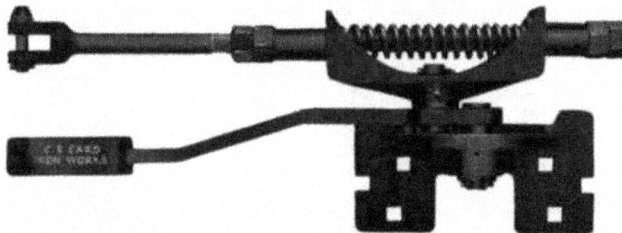

STAND WITH SPRING CONNECTING ROD.

Specifications; Prices and Average Weights in Pounds, Each.

Stand Number	Rail Size	With PLAIN Connecting Rod		With SPRING Connecting Rod	
		Price	Weight	Price	Weight
$3\frac{1}{2}''$ Throw A.M.C. Standard 1620–A	12 to 20 Lb.	$4.90	24	. . . .	
1620–B	25 to 48 Lb.	$7.90	43	$11.25	57
1620–C	50 to 60 Lb.	$10.30	79	$13.60	90
4" Throw 978	12 to 20 Lb.	$4.90	24	. . . .	
863	25 to 48 Lb.	$7.90	43	$11.25	57
5" Throw 922	50 to 60 Lb.	$10.30	79	$13.60	90

"CARD" PARALLEL GROUND THROW SWITCH STANDS

A parallel throw stand that finds favor with the mining and industrial trade because of its strength, few working parts, and space required for installation and operation.

The stand is made of steel throughout excepting the cast iron lever weight.

All bearings and working parts are machine finished for smooth and continued operation. The crank, that actuates the connecting rod, has ample bearing both top and bottom. The stand is self-cleaning.

The steel plate top covers and protects the principal working parts from damage.

The throw is adjustable from 3" to 4" through the screw crank.

It is a low stand, the largest size extending only 4" above the top of the ties, enough to keep the working parts out of the dirt.

The steel connecting rods have an adjustable end. Through this end or the throw screw crank, or both, the switch points can be easily and quickly adjusted.

The spring connecting rod prevents damage to either stand or switch parts when cars trail through the switch with the stand set in the wrong direction.

Please specify "plain" or "spring" connecting rods when ordering. Unless specified we will furnish the plain connecting rod.

Carried in stock for immediate shipment.

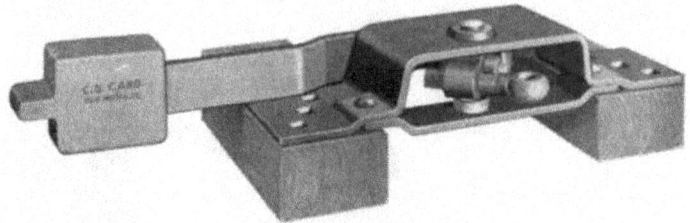

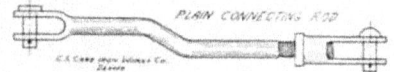

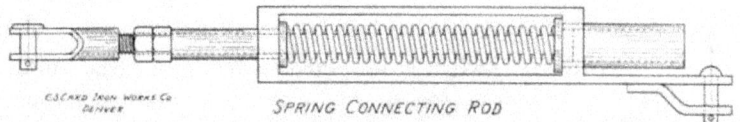

Specifications; Prices and Average Weights in Pounds, Each

STAND NUMBER	Rail Size	With PLAIN Connecting Rod		With SPRING Connecting Rod	
		Price	Weight	Price	Weight
PGT–20	12 to 25 lb.	$11.50	33	$16.30	56
PGT–30	30 to 48 lb.	$13.90	38	$18.90	64
PGT–40	50 to 60 lb.	$16.60	72	$22.15	102

Standard plain connecting rods are 16" and spring connecting rods 33" long center to center of connecting bolt holes.

"CARD" SPRING HEAD RODS

Spring head rods can be used in place of plain head rods. They prevent damage to either stand or switch parts when cars trail through the switch with the stand set in the wrong direction.

STYLE "A"

Style "A" rods are for use with 30" and wider gauges.

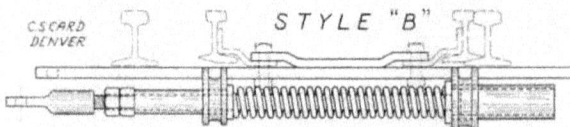

STYLE "B"

Style "B" rods are for use with less than 30" gauge and set lower between the ties.

When ordering, please specify style, track gauge, and weight of rail with which the rod is used.

"CARD" MALONE SPRING SWITCH STANDS

The Malone Switch Stand is the practical development of an ordinary spring switch, a positive switch car actuated, or ground throw, combining the good qualities of all in one device.

The switch points can be thrown by hand or automatically thrown by the car, and the stand is so constructed that it holds the switch rigidly, either in the closed or open position.

It is simple in construction and positive in its action, having two springs which hold the points rigid. It will be found a great convenience in all classes of mining and industrial railways.

The manufacture of the Malone Switch Stand is controlled exclusively by the C. S. Card Iron Works Co.

Unless specified, connecting rods will not be furnished.

Carried in stock for immediate shipment.

C S CARD
IRON WORKS Co
PAT. NO. 970343

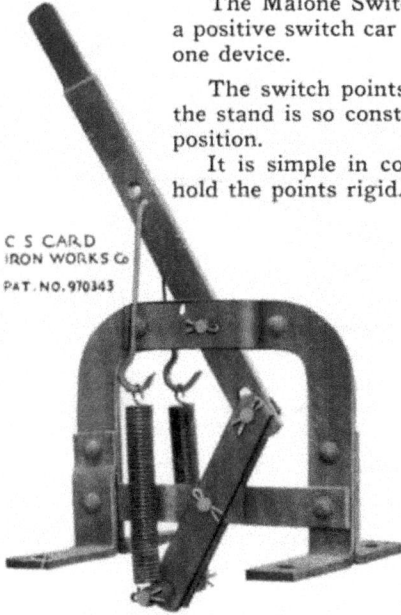

Specifications; Prices and Average Weights in Pounds, Each

STAND NUMBER	Switch Stand without connecting rod				*Connecting Rod	
	Rail Size	Maximum Throw	Price	Weight	Price	Weight
2	12 to 35 Lb.	4¼"	$10.00	45	$1.60	9
3	30 to 48 Lb.	4½"	$11.10	58	$1.75	12
4	50 to 60 Lb.	5¼"	$12.80	80	$1.75	12

*Standard Connecting Rods are 24" long center to center of connecting bolt holes.

"CARD" PARALLEL THROW SPRING SWITCH STANDS

This stand has all the advantages of the one above and with the addition of the parallel throw feature.

By using these stands the splitting of switches is practically eliminated as the switch points are always held tight against the rail.

Unless specified, connecting rods will not be furnished.

Carried in stock for immediate shipment.

Specifications; Prices and Average Weights in Pounds, Each

STAND NUMBER	Switch Stand without connecting rod				*Connecting Rod	
	Rail Size	Maximum Throw	Price	Weight	Price	Weight
12–PS	12 to 35 Lb.	4¼"	$19.25	66	$0.75	3
13–PS	30 to 48 Lb.	4½"	$20.70	82	$0.75	3
14–PS	50 to 60 Lb.	5¼"	$22.90	111	$0.80	4

*Standard Connecting Rods are 8" long center to center of connecting bolt holes.

Patent No. 1790795

"CARD" DIAMOND CROSS-OVERS

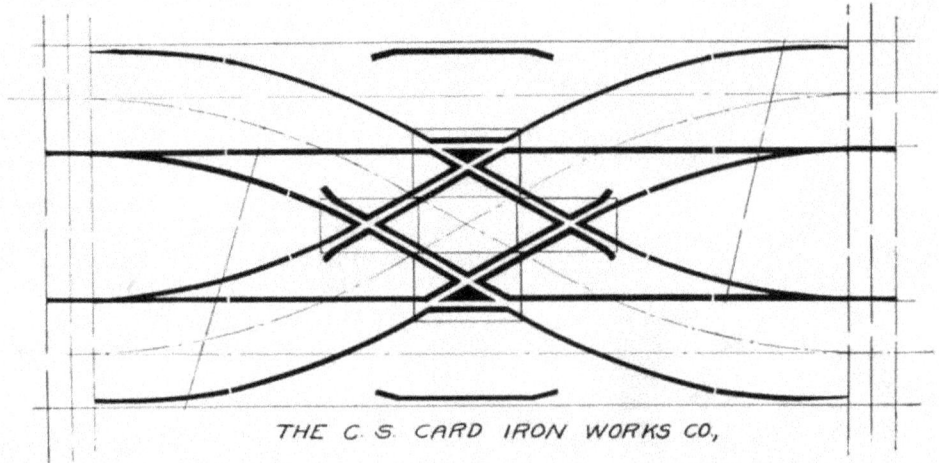

THE C. S. CARD IRON WORKS CO,

The illustration shows a Diamond Cross-over where the track centers are greater than the track gauge. We have designed and furnished a number of these for shaft bottoms.

Prices quoted on receipt of specifications. Advise us the weight rail, track gauge, center to center of tracks, and the maximum length for installing.

"CARD" CROSSINGS

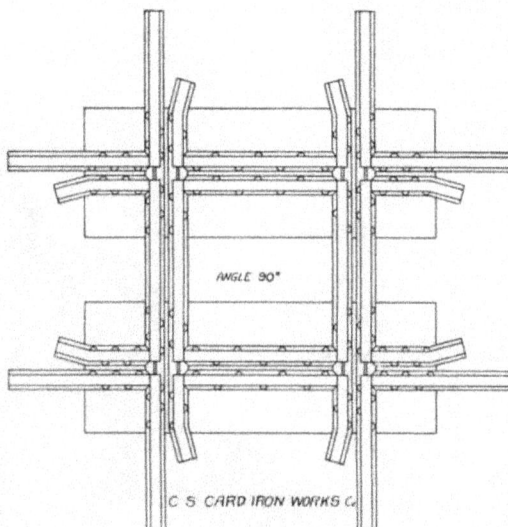

ANGLE 90°

C S. CARD IRON WORKS C.

The illustration shows one style of Crossing with guard rails, substantially attached to steel plate, that will be found very satisfactory for heavy travel. We make several styles of crossings. Prices quoted on application. Advise the weight rail, track gauge and angle of crossing.

"CARD" INCLINE LAYOUTS

Two to Three Rail and Three to Four Rail Incline Tracks
THE C S CARD IRON WORKS CO

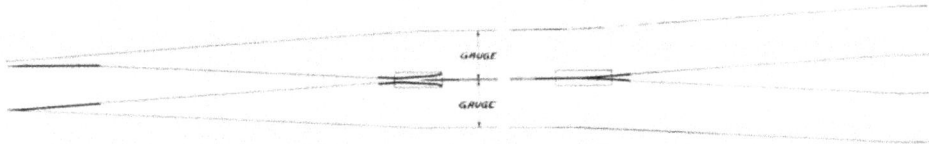

Single Point Rail Incline Frog

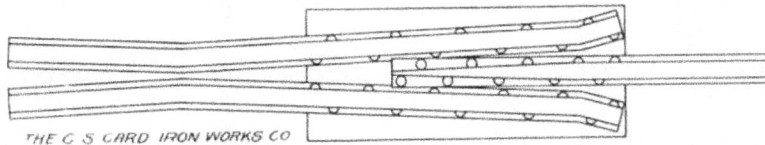

THE C S CARD IRON WORKS CO

Two Point Rail Incline Frog

THE C S CARD IRON WORKS CO

Illustrating the parts for the ordinary Two to Three Rail and Three to Four Rail Incline tracks. Prices quoted on receipt of specifications. Advise weight of rail and track gauge.

"CARD" INCLINE AND SHAFT BOTTOM LAYOUTS

We design and furnish trackwork to meet all kinds of shaft bottom and incline conditions.

Illustrated is a two to three track layout.

Prices quoted on receipt of specifications.

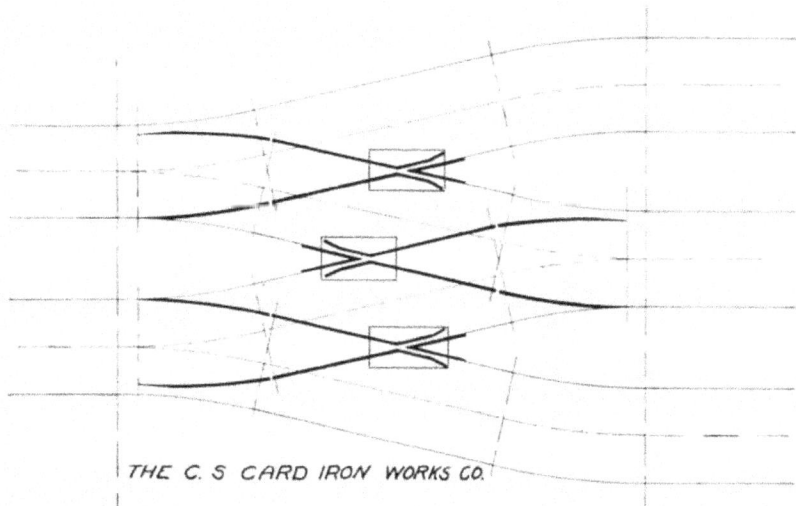

THE C. S CARD IRON WORKS CO.

LIGHT RAIL SECTIONS

The outline drawings and the table below show the standard dimensions and drilling for A. R. A. (American Railway Assn.) light rail sections, 8 lbs. to 48 lbs. per yard; and A. S. C. E. (American Society Civil Engineers) rail sections, 50 lbs. to 60 lbs. per yard.

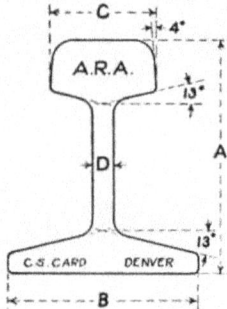

We will fill all orders for rail work made from these rail sections, drilled as shown, unless you specify a special section and drilling.

Heavier rail sections than shown in the table can be furnished.

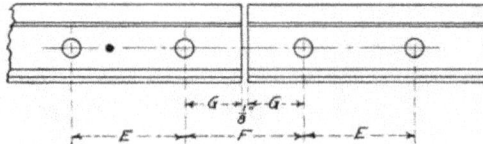

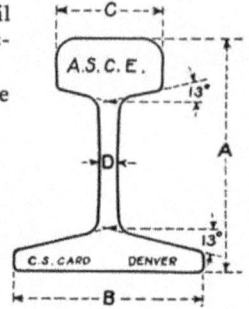

	Weight of Rail Per Yard	RAIL				DRILLING OF RAIL			
		A Rail Height	B Flange Width	C Head Width	D Web Thickness	E	F	G	Diameter of Holes
A. R. A. Rails	8 lbs.	1 9/16"	1 9/16"	25/32"	5/32"	4"	4 1/8"	2"	1/2"
	12 "	2 3/64"	1 3/4"	59/64"	3/16"	4"	4 1/8"	2"	5/8"
	14 "	2 3/64"	1 27/32"	1 1/64"	9/32"	4"	4 1/8"	2"	5/8"
	16 "	2 7/16"	2"	1 7/64"	7/32"	4"	4 1/8"	2"	5/8"
	18 "	2 7/16"	2 5/64"	1 3/16"	19/64"	4"	4 1/8"	2"	5/8"
	20 "	2 13/64"	2 1/4"	1 17/64"	1/4"	4"	4 1/8"	2"	5/8"
	25 "	2 13/16"	2 1/2"	1 29/64"	9/32"	4"	4 1/8"	2"	3/4"
	30 "	3 11/64"	2 3/4"	1 19/32"	5/16"	4"	4 1/8"	2"	7/8"
	35 "	3 23/64"	3"	1 23/32"	11/32"	4"	4 1/8"	2"	7/8"
	40 "	3 35/64"	3 3/16"	1 53/64"	3/8"	5"	5 1/8"	2 1/2"	7/8"
	45 "	3 47/64"	3 3/8"	1 63/64"	13/32"	5"	5 1/8"	2 1/2"	7/8"
	48 "	3 47/64"	3 29/64"	2 1/16"	31/64"	5"	5 1/8"	2 1/2"	7/8"
A.S.C.E. Rails	50 "	3 7/8"	3 7/8"	2 1/8"	7/16"	5"	5 1/8"	2 1/2"	1"
	55 "	4 1/16"	4 1/16"	2 1/4"	15/32"	5"	5 1/8"	2 1/2"	1"
	60 "	4 1/4"	4 1/4"	2 3/8"	31/64"	5"	5 1/8"	2 1/2"	1"

RAIL BENDERS

The illustration shows a Jim Crow Rail Bender. No description is needed, as all track men are familiar with it.

Made from the best grade forging steel, with machine cut square thread steel screw. Furnished without steel lever unless specified.

Specifications; Prices and Average Weights in Pounds, Each

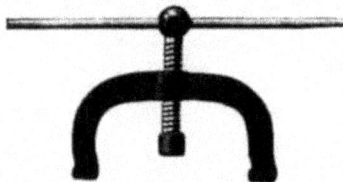

Size Number	Capacity Size Rail	Span of Hooks	Diameter of Screw	Weight	Price Without Lever Bar
6	16 Lb.	14"	1 3/4"	40	$15.00
5	25 Lb.	16"	2"	65	$19.00
4	50 Lb.	20"	2 1/4"	85	$25.00
3	75 Lb.	24"	2 1/2"	110	$29.00

"CARD" GUARD RAILS

Style "A" is furnished with the flares bent and with two cast iron chocks and bolts. The flange is not removed on the lighter weight rails.

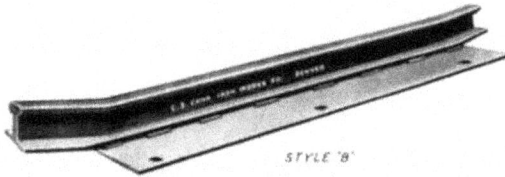

STYLE "A"

STYLE "B"

Style "B" guard rails are recommended. The rail is mounted on a plate, which is slipped under the main rail and spiked in position. There are no loose parts.

In ordering guard rails, specify the weight of the rail and the length overall. Standard lengths are 3'-0", 4'-0", 5'-0", 6'-0" and 7'-6".

"CARD" RISER SLIDE PLATES WITH RAIL BRACE

This combination plate and rail brace is recommended at switches where there is heavy traffic. It helps hold the track to gauge. Can be furnished for use with any of our standard switches, for any rail section.

"CARD" COMPROMISE RAIL JOINTS WITH STEP CHAIRS

Used for joining two different sizes of rail. Standard lengths are 36" long. Any combination of rail sizes can be furnished.

The step chairs can be omitted if desired, but this is not recommended since the smallest rail will not have the proper support and trouble will likely follow.

In ordering, please refer to "Rail Section" and "Holes for Fish Plates" on page 96. Be sure to furnish this information for both rails.

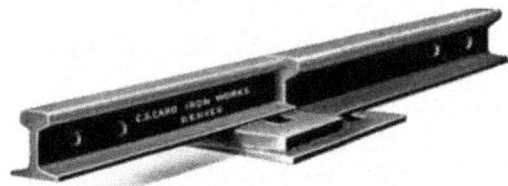

"CARD" BALL-BEARING TURNTABLES

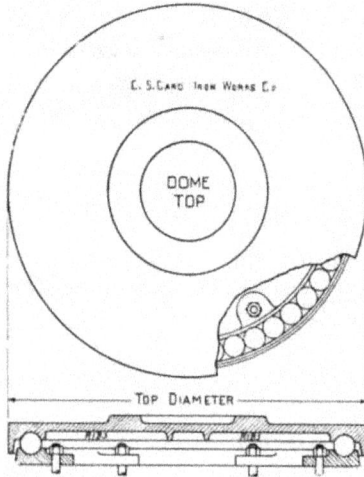

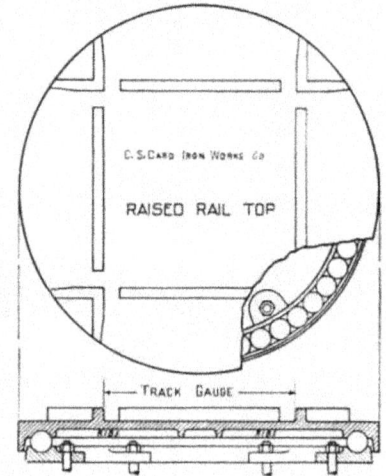

These turntables can be used to advantage in and about mines, quarries and industrial plants, where the use of a switch or switches is impractical.

Our tables are designed with the balls located as closely as possible to the outside, which gives stability, especially when the car is run on the table.

The downward projection on the outside edge of the table strengthens it and helps prevent dirt from entering the race.

These turntables are furnished with either smooth or corrugated flat top, dome top, or raised rail top.

In ordering always specify the style of the table top and the track gauge for dome and raised rail tops.

Specifications; Prices and Average Weights in Pounds, Each

Top Diameter	Maximum Wheel Base Recommended					Price	Weight
	18″ Gauge	24″ Gauge	30″ Gauge	36″ Gauge	42″ Gauge		
34″	18″					$41.00	481
40″	26″	20″				$55.00	615
48″	34″	30″	24″			$67.50	835
60″	46″	42″	38″	32″	26″	$110.00	1156

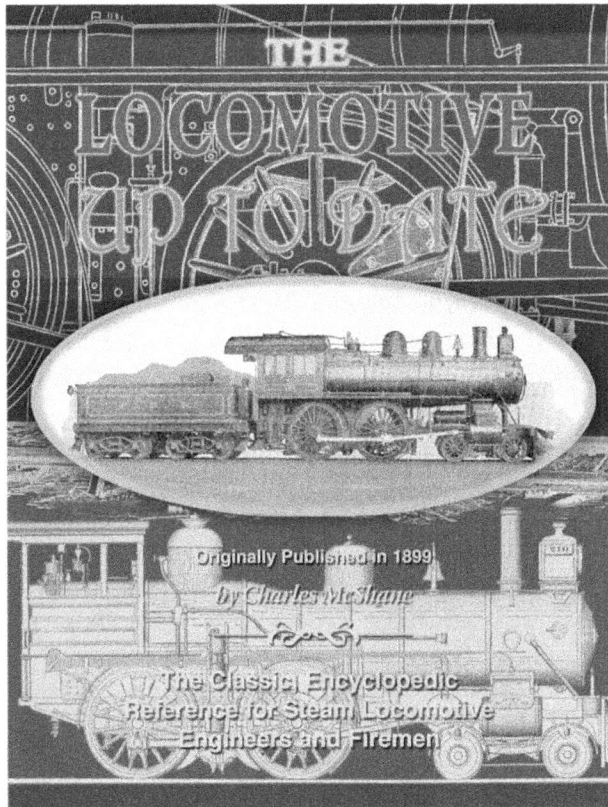

THE CLASSIC 1911 TROLLEY CAR BUILDER'S REFERENCE BOOK

ELECTRIC RAILWAY DICTIONARY

By Rodney Hitt

Associate Editor, Electric Railway Journal

ELECTRIC RAILWAY ENGINEERING

By Francis H. Doane, A.M.B.

www.ingramcontent.com/pod-product-compliance
Lightning Source LLC
Chambersburg PA
CBHW081233090426
42738CB00016B/3279